Topical Conference: Next-Gen Manufacturing

Held at the 2024 AIChE Annual Meeting

San Diego, California, USA
27-31 October 2024

ISBN: 979-8-3313-1685-3

Printed from e-media with permission by:

Curran Associates, Inc.
57 Morehouse Lane
Red Hook, NY 12571

Some format issues inherent in the e-media version may also appear in this print version.

AIChE
120 Wall Street, FL 23
New York, NY 10005-4020

Phone: (800) 242-4363
Fax: (203) 775-5177

www.aiche.org

Additional copies of this publication are available from:

Curran Associates, Inc.
57 Morehouse Lane
Red Hook, NY 12571 USA
Phone: 845-758-0400
Fax: 845-758-2633
Email: curran@proceedings.com
Web: www.proceedings.com

TABLE OF CONTENTS

3D PRINTING FUNDAMENTALS AND APPLICATIONS

3D PRINTING NOVEL METHODS AND APPLICATIONS

3D PRINTING OF COMPOSITES

3D PRINTING OF FUNCTIONAL MATERIALS

ADVANCES IN SMART MONITORING, OPTIMIZATION AND CONTROL OF PROCESS MANUFACTURING

APPLIED ARTIFICIAL INTELLIGENCE, BIG DATA, AND DATA ANALYTICS METHODS FOR NEXT-GEN MANUFACTURING EFFICIENCY

CYBERSECURITY AND HIGH-PERFORMANCE COMPUTING IN NEXT-GEN MANUFACTURING

FUTURE OF MANUFACTURING AND EMERGING TECHNOLOGIES

INDUSTRY 4.0, DIGITAL TWINS, AND DIGITAL TRANSFORMATION

MODELING, OPTIMIZATION, AND CONTROL IN NEXT-GEN MANUFACTURING

NEXT-GEN MANUFACTURING IN CHEMICAL AND ENERGY SYSTEMS

NEXT-GEN MANUFACTURING IN PHARMA, FOOD, AND BIOPROCESSING

POLYMERS IN ADDITIVE MANUFACTURING I

POLYMERS IN ADDITIVE MANUFACTURING II

440a *Dynamic 3D Printing of Structure Color*

Ying Diao

Biological systems have evolved to exhibit dynamic, hierarchical structures that confer complex functionality, such as the adaptable, structural color in chameleons that allows them to match their environment. Attaining such dynamic complex structures at the nanoscale have been challenging to achieve in synthetic macromolecular systems. In addition, synthetic colors we use today constitutes one of the biggest environmental pollutants that severely impact human and aquatic life. Developing printable dynamic structure color will not only introduce new functionality that current synthetic color is not capable of, but also help address this urgent environmental issue. Towards this aim, we and collaborators developed a programmable 3D printing process that can modulate nanoscale assembly and structure color of bottle brush block copolymers on the fly. By understanding and controlling the polymer assembly pathways during processing, we demonstrated 3D printing structure color as chameleon patterns without changing the ink material. We and collaborators further unveiled the underlying mechanism by probing polymer assembly pathways via a combination of X-ray scattering, optical spectroscopy, electron microscopy and coarse-grained simulations – the programmable structure color was attained through arresting polymer chain extension and entrapping metastable structures by controlling the assembly kinetics during printing.

440b Fiber-Based Hydrogels As 3D Printable Granular Materials with Unique Properties for Addressing Challenges in Biofabrication

Christopher B. Highley, Greg Grewal, Jack Whitewolf

Particle-based (granular) hydrogels are dynamic systems that can transition from a solid- to a liquid-like state under applied stress and back, upon the removal of stress, in the absence of interparticle crosslinking. As a function of this property alone, these materials have emerged as attractive (bio)inks and support baths in state-of-the-art 3D bioprinting processes. While these materials are typically formed from spherical particles, next-generation systems with new potential for engineering functionality are emerging in the form of fiber-based granular hydrogels. Here, we present granular hydrogels formed either partially or completely from discrete hydrogel microfibers that are ~1-2 microns in diameter but that can be 100s of microns in length. Compared to spherical particles, nanofiber particles' aspect ratios result in long-range interparticle interactions that give rise to new viscoelastic properties. Additionally, nanofibers can support high porosity in granular materials. We have created hydrogel nanofibers from both hyaluronic acid and poly(ethylene glycol) backbones. After electrospinning, fibers were dry-crosslinked, hydrated, and then segmented by homogenization to form the discrete microfiber particles. These were suspended in aqueous medium with or without cells, and then centrifuged to concentrate the hydrogel microfibers (and cells, if present) into macroscale granular hydrogels. Rheological and extensibility testing showed that, like traditional granular hydrogels, microfiber-based hydrogels exhibited a yield stress and can transition between solid- and liquid-like behaviors depending on applied stresses. Unlike traditional particle-based hydrogels, however, they were highly extensible (stretching to > 2000% bulk length). These materials also exhibited rapid stress relaxation, on the order of 10s of seconds, that can be controlled in part through formulation. In extrusion-based applications, such as bioprinting, the long-range entanglements stabilize printed filaments, even absent interfiber crosslinking. Interparticle crosslinking can be introduced within these systems to stabilize 3D structures. By introducing fibers into highly-porous granular hydrogels based on spherical particles, long-range fiber-mediated crosslinking can stabilize granular systems that would otherwise degrade by erosion. The new material functionalities enabled by these systems enable cell behaviors, such as spreading or the formation of microvascular network-like structures, without the need to design the hydrogel chemistry to support for degradation or yielding. We have seen that fiber-based granular materials support embedded printing, including allowing channels to formed by the removal of fugitive inks. Taken together, we believe fiber-based granular systems will present new opportunities for designing 3D materials for biomedical applications.

440c 3D Printed Biosponge Polymer Adsorber for Capturing Hazardous Chemotherapy Drugs to Reduce Chemotherapy Toxicities in the Body

Hee Jeung Oh

3D printing of multifunctional polymers can form previously nonexistent 3D complex structures that are difficult to prepare via conventional polymer processing. Integration of polymer chemistry with 3D printing leads to new polymer structures, opening a new door for biomedical separation. Here, we have designed, built, and deployed porous biosponge adsorbers for capturing chemotherapy drugs from the blood stream after these drugs have had their effect on a tumor, but before they are spread through the body where they can cause hazardous side effects. The support structure of the adsorbers was built using the continuous liquid interface production (CLIP). This structure was coated with a nanostructured block copolymer with outer blocks that anchor the polymer chains to the 3D printed support structure and a middle block that has an affinity for the drug. The middle block is polystyrenesulfonate which binds to doxorubicin, a widely used and effective chemotherapy drug with significant toxic side effects. The adsorbers are designed for deployment during chemotherapy using minimally invasive image-guided endovascular surgical procedures. We show that the introduction of the adsorbers into the blood of swine models enables the capture of 64 ± 6 % of the administered drug (doxorubicin) without any immediate adverse effects. This development represents a significant step forward in minimizing toxic side effects of chemotherapy.

440d Adaptive, in-Situ 3D Printing Using Multiphase Microfluidic Control Enables Multi-Materials Integration

Guillermo Ramirez, Gabriel Garibaldi, Gongchen Sun

The development of 3D printing technologies has heralded a new era in microfabrication, enabling complex designs inaccessible to conventional soft lithography methods. Among various 3D printing techniques, Digital Light Processing (DLP) is the most popular method to fabricate micro-scale structures due to its capability to create high-resolution features (tens of micrometers) at a much lower cost. For example, DLP-based techniques have been reported to create microfluidic devices with complex 3D structures and fluid-actuating parts such as on-chip pumps and valves. However, the focus of most studies remains predominantly on the creation of mechanical structures using photo-curable resins. The integration of functional materials in DLP-type 3D printing, such as conductive polymers and functional hydrogels, is limited. This is because DLP 3D printing relies on the layer-by-layer manufacturing of a prime material in a resin bath. While multi-material printing methods exist, they typically require the frequent substitution of the primary photopolymer material—a process that is not only labor-intensive but also potentially detrimental to the print's quality, as each introduction of new material requires the prints to be cleansed with solvents. Additionally, the spatial integration of multiple materials is limited, with the embedding of functional materials within the existing void features particularly challenging.

To address this challenge, we report an innovative DLP-based 3D printing approach, termed adaptive in-situ 3D printing. This approach facilitates the dynamic modulation of the printing layer in the third dimension, the Z-direction, within a microfluidic channel, leveraging the distinct advantages of multiphase laminar flow. Instead of using the VAT and build plate to create the printing layer, we use flow focusing of a photo-curable resin pre-polymer stream in a microchannel to precisely define the printing layer. By modulating the thickness of the resin pre-polymer stream, we can dynamically adjust the height of the print layer based on the design to minimize the fabrication time and the unintended crosslinking. To ensure the precise pattern on the XY-plane, an innovative self-focusing step is incorporated, leveraging image-based feedback alongside precise z-axis control to fine-tune the focal printing plane. This is complemented by an automated flow system for exchanging printing reagents, facilitating the sequential DLP printing of diverse functional materials within the microfluidic channel. Our approach enables the formation of intricate three-dimensional structures embedded with a broad spectrum of photopolymerizable materials in void spaces—overcoming one of the significant challenges faced by conventional stereolithography and soft-lithography methods.

We demonstrate the utility of our technique by integrating an ion-exchange polyelectrolyte and conducting polymer electrode in a microfluidic channel to achieve a field-flow fractionation device for biomolecule extraction. The integration of materials directly on-chip using our 3D printing method significantly enhances the electric field's coupling efficiency within the microchannel and make it possible to drive the electrophoretic biomolecule separation with low-power batteries. We envision that our technique could open new avenues for sophisticated on-chip functionalities in microfluidic sensor and biomedical applications.

440e Investigating Glyphosate Toxicity through the Development of Gut Organoids and 3D Printed Thin Films

Jonathan Appiah, Nicholas Edenhoffer, Brandon Bell, Shay Soker, Mohammad Azad

Glyphosate (N-Phosphonomethylglycine), the most widely utilized *herbicide* in modern agriculture, has prompted concerns regarding its environmental impact and potential health implications to non-target organisms, particularly humans. Traditional toxicological methods provide controversial information on glyphosate toxicity and may not accurately reflect human responses, as 2D in-vitro models lack human physiological complexity. This research aims to develop a 3D micro-engineered organ equivalent (MOE) of the human small intestine (gut organoid), aiming to mimic the *in-vivo* environment more effectively than traditional 2D cell cultures. This gut organoid will be used to study the toxicity of glyphosate incorporated in a 3D-printed thin film matrix. CaCO2 and nHDF cells were cultured in modified Collagen I solution at different concentrations (0.5 and 2 mg/ml) and cell ratios (1:1, 1:10, and CaCO2 only, respectively) to develop gut organoids. Immunostaining analyses (IHC) were performed on Days 7 and 11 to evaluate epithelial (ZO-1) and stromal (vimentin) markers. To develop the 3D printed film, a 20-gram water-based ink was formulated using Hydroxypropyl Methylcellulose (HPMC) as base polymer, glyphosate, and other functional excipients (glycerin and polyvinylpyrrolidone). Rheological analysis was performed to characterize the ink properties, including flow/viscosity, amplitude sweeps, and thixotropy tests. Subsequently, 3D films were successfully printed from a custom design using a CELLINK BIOX printer using a print pressure of 60 kPa and a print speed of 10 mm/s. The gut organoid development suggests that the 2 mg/ml collagen concentration with both cell types produces more favorable epithelial morphology than the 0.5 mg/ml concentration. The presence of fibroblasts is essential for the formation of the epithelial layer could be deduced from the organoid formation. The non-uniform shapes of the organoids which posed challenges for preparation and sectioning, were further tuned through 3d printed well plates with structures to support the spherical formation of the organoids during crosslinking. The overall study shows that a gut organoid was fabricated, and 3D-printed glyphosate films were successfully printed. These platforms will allow us to conduct *in vitro* toxicological evaluation of glyphosate at concentrations relevant to human physiology. Future works to be concluded before the conference include the dissolution profile of the glyphosate films, as well as the permeability of the gut organoids using 1 mg/ml FITC Dextran fluorescein-isothiocyanate (FITC)-labeled Dextran 4 kDa. The impact of various relevant glyphosate concentrations exposure to the gut organoids will be evaluated through Live/Dead cytotoxicity, ATP assays, and disruption of the epithelial barrier.

440f Development of FeCo Catalysts Supported on TiO_2 Nanotube Arrays Grown on 3D-Printed Structures: Applications in Fischer-Tropsch Synthesis

Luis Caballero, Michael Nigra

Fischer-Tropsch Synthesis (FTS) is a prominent reaction for converting synthesis gas (syngas) into high-value liquid chemicals and fuels. Syngas is derived from the thermal conversion of coal, but it can also be obtained from biomass and natural gas. FTS yields olefins, paraffins, and oxygenated molecules (alcohols and carbonyls) via a series of surface polymerization reactions. However, its optimization remains challenging due to its complexity and dependence on process parameters, including temperature, pressure, residence time, and catalyst type. The support material in an FTS catalyst dictates CO conversion and C_{5+} selectivity, with TiO_2 remaining a popular option due to its abundance and desirable material properties. The development of state-of-the-art technologies in materials science further propels efforts to enhance TiO_2's functionality as a support. Additive manufacturing (3D printing) has gained prominence in the manufacturing of metal components due to its versatility and reduction of energy consumption.

In this contribution, 3D-printed Ti-6Al-4V structures are used as catalytic support substrates for developing superficial TiO_2 nanotube arrays (TNAs). This architecture is hypothesized to be desirable because it facilitates molecular diffusion and minimizes the complexity of its porous structure. These nanostructures are optimized by annealing, with crystal structure, morphology, and elemental composition changes evaluated via FESEM, XRF, XRD, and EDXS. Bimetallic catalysts are then synthesized by impregnating the TNAs with iron (Fe)-cobalt (Co) nanoparticles, and their activity is tested for FTS at 250 °C. The results show that the optimal sequence for the synthesis of TNAs involves a two-step electrochemical anodization in ammonium fluoride (NH_4F), followed by low-temperature annealing at 450 °C. The resulting catalysts are active for FTS, displaying a rate of CO consumption of $2.43 \times 10^{-5} \pm 8.99 \times 10^{-6}$ $mol_{CO}/s \cdot g_{FeCo}$ and a chain growth probability (α) of 0.273 ± 0.143.

294a Rheological Considerations for Materials and Process Design in Dense Suspension 3D Printing

Michael Bortner, John Reynolds, Christopher B. Williams, Daniel Rau

Additive manufacturing (AM) is revolutionizing manufacturing by enabling complex geometries and multimaterial structures that cannot be fabricated using traditional approaches. Extrusion additive manufacturing is one of the most identifiable modes of AM that implements predominantly polymeric materials. However, the technology is limited in terms of material availability and corresponding process design. Researchers have been pushed by consumers and industry to stretch the capabilities of AM across many different modalities.

Polymer extrusion AM, which encompasses modalities commonly referred to as pellet fed AM (BAAM), fused filament fabrication (FFF) and direct ink write (DIW) have differing and unique rheological requirements that can guide both materials and process design for enabling "printable" polymers and polymer composites. There is significant interest to enable AM of dense polymer matrix suspensions with high loading levels of rigid particles and fiber fillers to enable enhanced printability, functionality and mechanical performance. In this talk, I will focus on probing fundamental rheological considerations to enable polymer extrusion AM of opaque, high solids content systems in the DIW material extrusion platform, with an emphasis on dual-cure matrix technologies for DIW based processing.

294b Beyond Layers: 3D-Printable Morphing Architectures Via Programmable Stress

Duanduan Han, Victor M. Ugaz

Significant advancements in soft robotics have been achieved using polymeric materials that actively respond to external stimuli via shape memory or physico-chemical interactions with their surroundings. Recent excitement has been generated by applying these responsive properties to produce shape-shifting 2D structures capable of morphing into complex 3D topologies. However, manufacturing these active components generally requires specialized chemical formulations and sophisticated multi-material patterning capabilities only available in research laboratory settings, significantly limiting their impact. Here, we show how this barrier can be overcome by using 3D printing to embed prescribed internal stresses within a planar 2D substrate, mirroring the hierarchical structures found in living systems. We introduce a library of elemental building blocks designed to distribute strain across multiple scales and show how these components can be assembled into active systems with tunable mechanical properties. Finally, we apply these principles to produce functional devices such as self-activating oleophilic sponges, precision mechanical grippers, and autonomous self-tying knots. Importantly, these components can be manufactured using standard 3D printers and materials, thereby democratizing access to soft robotics technology.

294c Additive Manufacturing of Multi-Metal Structures with Adjustable Gradients Via Continuous Liquid Interface Production

Dylan Joralmon

Metal additive manufacturing (AM), also known as metal 3D printing, has become the go-to technology for fabrication of high-quality metal and alloy components due to their high integrity and customizability providing a method for manufacturing intricate designs with broad applications within the biomedical, aerospace, and military sectors. Well-established AM technologies such as powder bed fusion, material extrusion, and binder jetting have demonstrated the ability to manufacture novel designs in comparison to conventional casting and injection molding techniques that are expensive and wasteful. However, these metal AM technologies have limited material selection and anisotropic properties that stem from the layer-based printing approach that is inherent to these methodologies leading to non-uniform grain size distribution and "staircasing" effect which can lead to undesired performance and have a negative impact on the exhibited physical properties of the metal structure. Furthermore, when fabricating metal objects with two or more metals, the powder recoating and particle fusion process is constrained by long print times and expensive hardware since each printable powder material has distinct melting characteristics and thermal properties that require precise process control to ensure part integrity. Recent developments of low viscosity printing resins with metal precursors utilizing continuous liquid interface production (CLIP) 3D printing have relaxed the challenges associated with poor surface roughness and printing speed of meso- and microscale functional metal devices allowing for next-generation thermal management devices and metamaterials with isotropic performance. Moreover, the continuous AM technologies can be utilized to enable the fabrication of multi-metal components with composable gradients with tunable physical properties and print times that are orders of magnitude faster compared with traditional metal 3D printing. Consequently, the research presented here aims to address key performance parameters, including material deposition, printing performance of metal salt resins, thermal characteristics of multi-metal precursors, and post-processing settings, in order to develop a novel AM technology that can achieve multi-metal objects with homogeneous grain size distribution and tunable physical properties at higher resolutions. Single and alloy metal salt printing resin with low viscosity are synthesized utilizing inorganic metal salts enabling the fabrication material to be evacuated and refilled in a facile manner during the 3D printing process to achieve metal components with more than one metal. Experimental data on the post-processing parameters of multi-metal precursors indicate that both dense and porous multi-metal structures can be achieved via the mask-video projection vat photopolymerization with resin exchange process. Additionally, gradient metal components exhibit excellent surface quality, smooth material transition regions, and homogenous microstructure with uniform grain sizes on the top, bottom, and front faces. Lastly, the broad impact of this work is discussed, and a comparison is made between the microstructures and surface quality of multi-metal printed structures using the proposed approach with current state of the art metal AM technologies.

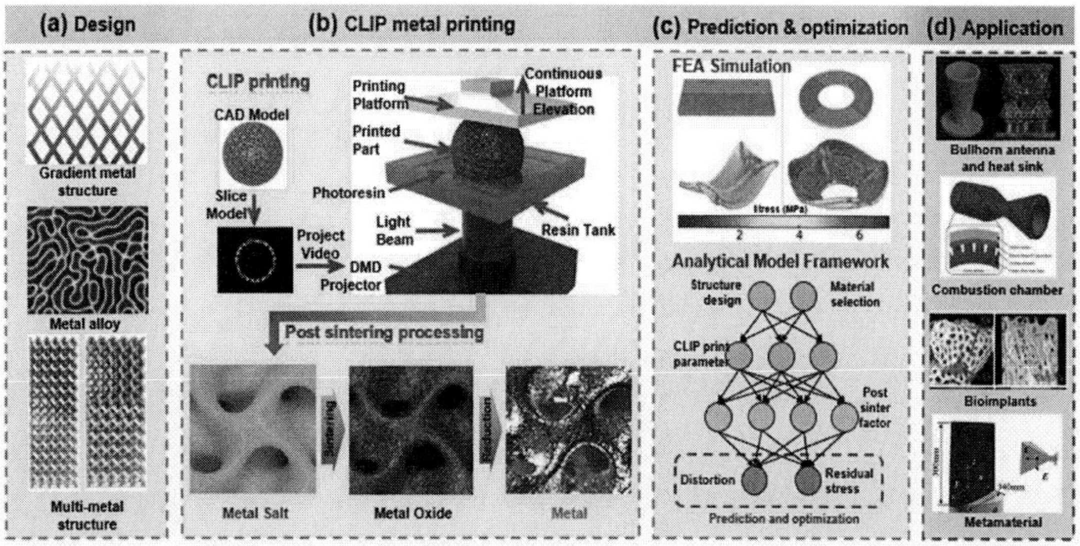

294e High-Throughput Optimization of Functional Materials Via Combinatorial Additive Manufacturing

Minxiang (Glenn) Zeng

The development and optimization of new materials are essential for the advancement of technologies in fields like renewable energy and environmental protection. The process of identifying and enhancing materials, however, has been traditionally slow and inefficient. Existing methods for combinatorial synthesis are capable of generating extensive libraries of materials, including chemical molecules and semiconductor materials, yet they are restricted by a narrow range of material options and fail to fully utilize the recent progress in colloidal nanomaterial synthesis. Here, we present a bottom-up combinatorial printing (CP) technique that facilitates the production of materials with gradient compositional variations for materials optimization. This technique allows for the on-the-fly adjustment of mixing ratios among a wide array of material choices. By altering the chemical composition of the inks and the dynamics of their deposition, we demonstrate a broad spectrum of printing techniques and high-throughput applications, including combinatorial doping, functional grading, and combinatorial reactions. As a proof of concept of CP, we explore the effects of doping levels on thermoelectric materials, efficiently determining the optimum doping level that leads to the synthesis of an n-type material with a superior power factor at room temperature. This method, blending the adaptability of additive manufacturing with precise control over the composition of precursor materials, presents a promising platform for expediting the synthesis and evaluation of a wide range of material systems.

References: Zeng, Minxiang, et al. 'High-throughput printing of combinatorial materials from aerosols.' Nature 617.7960 (2023): 292-298.

294f Micro to Nanometer Scale 3D Printing for Geometrical Design of Functionally Graded Polymeric Materials

Kaiwen Hsiao

Micro- and nano-meter geometrical structures dictate critical polymeric material properties ranging from mechanical robustness to transport selectivity. While additive manufacturing (AM) has enabled free-form patterning of polymeric materials, the trade-off in resolution and scalability has limited its capability to control material properties through simple geometrical designs. In this talk, we will discuss our development of single-digit resolution AM technology based on Continuous Liquid Interface Production (CLIP). Building upon photopolymerization kinetics, complex fluid dynamics, projection optics design, and software system integration, the single-digit micrometer CLIP 3D printer has enabled direct fabrication of micrometer scale, functionally graded, latticed capacitive pressure sensors with varying sensitivities for pressure and shear sensing. We found that by introducing different unit cell types and geometrical thicknesses, we can vary the sensitivities and thus the capacitive responses of the pressure sensors. An analytical model was established to capture the capacitive responses of pressure sensors that consist of different unit cell types, and a close match was obtained between the analytical results and experimental measurements. In-situ micro-CT scans were further conducted to capture the microstructure deformation under shear and normal force, and a qualitative finite element method was developed to capture the deformation of the micro-struts. Taken together, the single-digit micrometer resolution AM enables direct tuning of material mechanical properties through varying simple geometrical designs.

294g Binder Jet 3D Printing of Pea-Based Snacks with Tunable Texture and Precision Nutrition

Ethan Chadwick, Yara Suleiman, Guilherme Bertola, Sina Shahbazmohamadi, Abhishek Shetty, Anson W. K. Ma

To-date, material extrusion (MEX) is the most studied and published method for 3D printing food structures. In this presentation, we will share our latest findings in 3D printing snacks using the binder jetting (BJT) method, where a liquid binder is repeatedly deposited onto thin layers of food powder. Unlike MEX, this approach does not require the formulation of an extrudable paste which is more susceptible to shrinkage and cracking as the sample dries. Also, as only a small amount of water is needed, BJT is well suited for producing low-moisture food products. Using pea flour and an aqueous binder as feedstocks, we will elucidate the underlying binding mechanism and the effects of sugar inclusion and baking on the mechanical properties of the printed samples. The printed and then baked samples will be further compared with commercial snack bars that are not 3D printed. The mechanical properties of 3D printed samples are highly tunable through digitally programming the print head waveforms. In the second part of this presentation, we will present two different methods for precisely controlling the dosing of a micronutrient in the 3D printed snacks. All these results demonstrate the exciting potential of 3D printing for mass customization of snacks with tunable texture and precision nutrition. This work is supported by USDA National Institute of Food and Agriculture, AFRI project 2019–06721 (under Grant No. 2020- 67017-31273).

359a 3D Printing-Guided Chiral Nanostructures in Bio-Hybrid Nanocomposites

Monirosadat Sadati, Mohsen Esmaeili, Kyle George

The chiral or helical structures inherent to biomolecules contribute to the exceptional fracture resistance observed in the 'smasher-type' mantis shrimp's dactyl club and the vibrant metallic colors in beetles. Drawing inspiration from these natural helical structures, this study integrates 'bottom-up' molecular self-assembly with 'top-down' 3D printing to design intricate geometries with a guided, radially twisted chiral hierarchy using inks derived from cellulose nanocrystals (CNC). We utilize the processing parameters of direct ink-writing (DIW) 3D printing to manipulate the chiral nematic liquid crystalline phase in cellulose-based inks before deposition. This approach allows us to steer structural development from a specific initial condition. Through comprehensive rheological analyses, we have found that printing the chiral inks at a shear rate that induces a pseudonematic state in cellulose molecules leads to uniform chiral recovery after the cessation of flow. The subsequent biomineralization of these printed structures with a chiral arrangement of CNC particles has revealed significant enhancement in their mechanical properties, demonstrating a synergistic interaction between biomineralization and chiral assembly that induces nanoscale ordering. This biomimetic strategy offers a pathway to develop materials with superior mechanical properties derived from 3D nano/microstructures, which can be translated into larger scale 3D printed designs.

359b Design of Multiphasic Homocomposite Inks for Precise 3D Printing with Soft Materials

Orlin D. Velev

The use of composite inks for 3D printing allows precise adjustment of their pre- and post-extrusion characteristics but introduces complexity due to the inter-material interfaces. We will discuss the concept of "homocomposite" 3D printing via extrusion where both components are made of the same material but in different phases. Two engineering strategies which use multiphasic liquid-(bio)polymer systems for homocomposite 3D printing will be discussed. First, we will present 3D printing inks made by capillary binding, including new magnetically responsive gels and soft structures consisting of water, crosslinked PDMS microbeads and liquid PDMS phase. These Homocomposite Thixotropic Pastes (HTPs) enable the formation of remarkably elastic, flexible, and biocompatible silicone structures. The HTP-3DP inks open new opportunities for 3D printing of "active" and magnetically reconfigurable structures. In the second approach, we use homocomposite gel from alginate with double network, including molecular and colloidal networks. The colloidal network is made of soft dendritic colloids (SDCs) discovered and investigated earlier by our group. The SDC have high excluded volume and strong van der Waals interactions and readily form percolated 3D networks, resulting in gelation at low volume fractions. The alginate SDCs can be incorporated into molecular alginate hydrogel matrices to form hierarchical networks with mechanical properties resulting from synergistic combination of the molecular and colloidal networks. The alginate homocomposites have remarkable mechanical properties, allowing to 3D print and inject these hydrogels into architectures with increased toughness. These inks also open the way to "green" 3D printing with sustainable homocomposites.

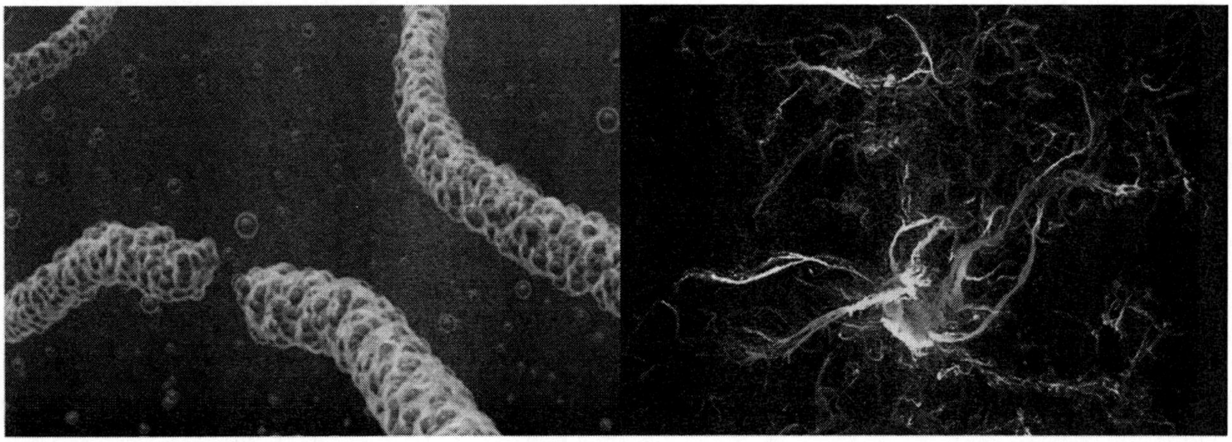

359c High Throughput Mechanical Analysis (HTMECH) of Fused Filament Fabrication (FFF) Printed Multi-Material Parts

Daniyal Shoukat, Javaz Rolle, Carson Meredith, Nese Orbey, Jay Park

Fused Filament Fabrication (FFF) is a promising 3D printing method of integrating the useful properties of two different materials into a single functional object through multi-material printing. The inefficient interlayer bonding between the two materials during printing, however, results in poor mechanical behavior of the end-part along with dimensional inconsistencies and delamination between layers, limiting the use of FFF for industrial applications. In addition, testing the tensile properties using conventional uniaxial tensile testing can be time consuming. In this work, High Throughput Mechanical Analysis (HTMECH) is developed as a rapid screening method to test the tensile properties and interfacial bonding of FFF printed multi-material parts in reduced time as a function of 3D printing conditions and material properties. This is done by comparing the tensile properties of multi-parts parts measured using conventional uniaxial tensile testing methods on dog-bones with ones obtained using HTMECH on dog-bones. Various compositions of polycarbonate (PC)-based resins are used to print neat filaments with reinforced filaments to analyze the effect of fillers (carbon-fibers, PC-ABS blend) on mechanical behavior, whereas Poly(Lactic Acid) (PLA) is printed with poly(caprolactone) (PCL) to evaluate the interfacial properties and mechanical behavior between two completely different materials.

359d Tuning Material Extrusion Additive Manufacturing for Water Remediation Applications

Michael Bortner, Allan Kennedy

Additive manufacturing (AM) is enabling complex geometries and multimaterial structures that cannot be fabricated using traditional approaches, opening novel design spaces for water treatment applications. Material extrusion additive manufacturing is one of the most identifiable modes of AM that implements predominantly polymeric materials, including broad ranges of filled systems. In this talk, I will highlight our efforts towards remediation of multiple water contaminants using particle loaded (up to >30 wt%) thermoplastic filament printing. Target contaminants in our work have included ammonia, polyaromatic hydrocarbons (PAHs), microcystin toxin released in harmful algal blooms (HABs) and perfluoroalkyl and polyfluoroalkyl substances (PFAS). The benefits afforded by additive manufacturing over traditional filled polymer processing methodologies and compared to current remediation strategies will be highlighted, including aspects such as deployability, selective local photocatalytic efficacy and targeted adsorption kinetics. The results presented here will also further challenge the traditional interpretation of "printability" and the application specific relevance of this concept.

359f Fused Filament Fabrication of Conducting Nanocomposites Based on Reduced Graphene Oxide/Polymer Blends

S Michael Kilbey II, Evan Holt, William Ledford

Adding functionality to 3D printed specimens presents a significant opportunity to benefit from the advantages of additive manufacturing while expanding utility of printed parts beyond structural components. Here, conducting composite filaments consisting of a blend of thermoplastics, including poly(methyl methacrylate) and other engineering copolymers, and reduced graphene oxide (rGO) were created via a solution casting process and used in fused filament fabrication. To promote intimate mixing, graphene oxide was reduced in a solution containing the matrix polymer and subsequently coprecipitated to produce a well-dispersed nanocomposite material. This strategy provides the capacity to vary the amount of the rGO additive incorporated in the blend and significantly change the electrical conductivity of the materials. Assessments of printing by fused filament fabrication demonstrate that the brittle nature of filaments necessitate the addition of polymeric plasticizers that enhance printability by altering rheological properties and compatibility. Ultimately, printed parts show increases in thermal and electrical conductivity with increasing rGO loading level, as well as changes in mechanical properties associated with the incorporation of the nanofiller rGO. In total, the results demonstrate the utility of the solution casting method as well as the opportunities to provide functionality based on the current-carrying capacity of the FFF-printed parts.

102a Tailoring Material Properties Using Additive Manufacturing Techniques

Anesia Auguste

Direct ink write (DIW) is a commonly used technique in polymer additive manufacturing, where filaments of material are deposited to build 3D structures. The development of desktop robotic arms have enabled the capability for conformal and support free printing within the DIW field; However, the research on the effects of printing with a robotic arm vs a gantry system on the material properties is limited. We investigated the effects of DIW with a robotic arm on the printability and material properties of an anisotropic material as compared to a gantry system. Using a carbon fiber epoxy composite ink, samples were mechanically tested, 3D scanned to investigate geometrical accuracies, and CT scanned to analyze fiber alignment to elucidate the role of nozzle orientation, kinematics, and gravity on the material properties. This research will provide insight into deposition technique effects on material properties, providing a better understanding of printing onto complex surfaces. Then, we will present a methodology to realize surface-templated homeotropic orientation in liquid crystal elastomers and omnidirectional nonlinearity in mechanical deformation. Liquid crystalline elastomers (LCEs) are widely recognized for their exceptional promise as actuating materials that can be utilized in aerospace applications, optics, or medicine. By inkjet printing the homeotropic alignment surface, we can localize regions of homeotropic and planar orientation within a monolithic LCE element. The local control of the self-assembly and orientation of the LCE yields materials with discontinuous mechanical deformation but with the advantage of lacking material–material interfaces, a common source of failure. This allows us the ability to create designer functional monoliths with tunable Poisson's ratios which may find use in many applications such as flexible hybrid devices.

102b 3D Printing of Surface Structure Reinforced Poly(ethylene glycol) Diacrylate (PEGDA) Electrolytes for All-Solid-State Li-S Batteries

Alexander Hoffmann, Wissam Fawaz, Simon Ng

Solid-state batteries are a key step in producing the next generation of high energy storage systems. Currently, all-solid-state-batteries (ASSB) are burdened with diminished ionic conductivity and poor interfacial contact between the solid-state electrolyte and the electrode materials leading to higher resistances and poor electrochemical performance. 3D printing has emerged as a promising solution to the poor interfacial contact between electrolyte and electrode materials, owed to the intricacies the technique affords to the engineering of surfaces. Specifically, 3D printing is capable of developing precise structures at the milli, micro, and now nano scale. By using advanced stereolithography, solid poly(ethylene glycol) diacrylate (PEGDA) based polymer electrolytes can be designed with different surface structures (i.e. concentric circles, parallel ridges, waffle structures, among others) that increase the interfacial contact area between the polymer matrix and electrodes. Employed novel surface structures on polymer electrolytes as a host for cathode materials allowed for improved interfacial contact and increased loading of active materials. The cathode/electrolyte interface will be characterized by SEM and TEM to gain a better understanding of the composition at the interfacial layer pre and post cycling. Assembled cells using 3D-printed electrolytes show improved room temperature cyclability and specific capacities in Li-S batteries. The findings of this work will gain a better understanding of 3D printed structures on the electrochemical performance, and lead to a rational design of solid electrolytes for the next generation of Li-S batteries.

102c Precision Formation and Stabilization of Photopolymerizable Microbeads for Enhanced Fluorescence-Based Thermometry in Microfluidics

Isteaque Ahmed, Shihab Mahmud Bhuiyan, Aashish Priye

This work develops a photocurable resin co-flow microfluidic system for the precision formation of multifunctional double emulsion droplets. Utilizing a PDMS-glass chip fabricated through 3D printed molds, we generate droplets containing a photocurable, biocompatible resin. This resin, a blend of polyethylene glycol diacrylate (PEGDA) and ethylene glycol polyether acrylate (EGPEA), allows for the in-situ solidification of droplets under UV illumination, thereby stabilizing their size and shape. The polymer composition of EGPEA and PEGDA was meticulously controlled, to modulate crosslinking kinetics. This enabled the successful encapsulation of Rhodamine B within the resin droplets to leverage its temperature-dependent quantum yield for non-contact thermometry applications. The emission peaks of Rhodamine B at 570nm, upon excitation at 546nm, remain distinct from the UV curing wavelength of the resin, ensuring the stability of the thermal reporting mechanism.

Through comprehensive experimental and numerical analyses, including constructing resin droplet flow regime maps and leveraging CFD simulations, the work offers novel insights into the physics of resin droplet generation. These insights encompass the effects of surface wettability, flow rates, and capillary numbers on droplet formation dynamics and stability within high-viscosity photopolymer systems. The research addresses critical gaps in current methodologies by presenting a scalable, efficient strategy for generating monodisperse, functionalized photocurable microbeads. Such advancements promise significant impacts on fields ranging from biosensing to precision medicine, highlighting the potential of microfluidics in revolutionizing microscale analysis and synthesis platforms.

102d Spinning Desicurer: A Cost Effective and Generalizable Post Processing Method for Enhanced Optical Quality in 3D Printed Microfluidics

Gabriel Garibaldi, Guillermo Ramirez, Miguel Garibaldi, Gongchen Sun

Microfluidics are miniaturized devices used for handling small amounts of fluids at the microscale and have been used for various applications, including environmental analysis, medical diagnostics, *in vitro* modeling of organs, and image-based screening. 3D printing, has paved the way to a new horizon for microfluidics by enabling rapid device prototyping, offering unprecedent flexibility in device design and minimal requirement of specialized equipment and expertise. Among various 3D printing approaches, stereolithography (SLA) 3D printing is the most popular to fabricate microfluidic devices due to the high spatial resolution it offers. SLA 3D printing uses patterned light to selectively crosslink photo-curable resin to achieve the device structure. Typically, post processing after printing is necessary, which includes a solvent wash step and an additional UV curing step to clean and strengthen the printed object. However, an often overlooked but critical aspect is that the porous nature of photo-crosslinked resin materials can result in the penetration and entrapment of these solvents in the print when alcohol-based washing solutions and resin-dissolving agents are used excessively. This results in a cloudy, translucent appearance of the printed device. Additionally, the lack of transparency can also be attributed to non-smooth printing surfaces of the build plate and the use of support structures. Lastly, the presence of oxygen during the post processing step can cause complication and contribute to the optical imperfection of the print surface since molecular oxygen is known to inhibit radical induced polymerizations due to its high reactivity toward radical species.

To address this challenge, we develop an integrated design to enable a simple, cost-effective, and generalizable post-processing method that removes the excess resin without using solvents and produces optically transparent microfluidic devices made by SLA 3D printing. Our equipment, termed "Spinning Desicurer" (short for "Spinning Desiccating Curer"), integrates a low-cost spinning motorized stage and a 395nm UV LED light in a vacuum desiccator. The Spinning Desicurer allows us to simultaneously expel the excess resin and create a smooth surface finish using centrifugal forces; At the same time, the device is UV-cured under a low-oxygen environment. Additionally, we introduce a simple build plate modification method using a flexible thin polyimide film to remove the need of supporting structure during printing and reduce surface imperfections. We quantify the transparency of the print using a Scale-Invariant Feature Transform algorithm on a reference image, which shows the print transparency made by our method is orders of magnitude higher than the traditional post processing method. We further demonstrated the optical quality of the microfluidic device made by our method with imaging of microflow and microparticles under different imaging modalities. The high-resolution images obtained via our device show high fidelity in both bright-field and fluorescence modes compared with the blurry and distorted images obtained from traditionally made devices. The optical property of our 3D-printed device is sufficient in most high-resolution image-based microfluidic applications such as phenotype-driven screening of cells and microorganisms. Our method can be executed using low-cost, easily accessible materials, circumventing expensive commercial apparatus, sophisticated printer modifications, and potential safety hazard of handling waste solvent. We hence envision that this novel method will be easily adopted in different labs and enable users to prototype microfluidic devices for high-resolution applications including but not limit to bioengineering, medical, and pharmaceutical fields, with minimal requirement in supplies and effort.

102e 3D Printable Carbon/Carbon Composites

Dharneedar Ravichandran, Yuxiang Zhu, Varunkumar Thippanna, Kenan Song, Arunachalam Ramanathan

This project introduces an innovative approach to manufacturing carbon-carbon (C/C) composites via 3D printing, intertwining artificial intelligence/machine learning (AI/ML) and comprehensive life cycle and techno-economic analyses (LCA/TEA) to optimize production processes. Led by Prof. Kenan Song, the team combines expertise in materials science, chemical engineering, mechanical engineering, and AI to revolutionize C/C composite fabrication. The initiative aims to enhance the structural, thermal, and mechanical properties of C/C composites, making them viable for high-performance applications in aerospace and automotive sectors. Through advanced 3D printing techniques, the project addresses the scalability and efficiency challenges in traditional C/C composite manufacturing, offering a path towards sustainable and economically viable production methods.

102f 3D Printable Biodegradable Polymers for Medical Applications

Kenan Song, Yuxiang Zhu, Dharneedar Ravichandran

This study presents the design and fabrication of multi-channel nozzles using 3D printing technology, which were utilized to guide biocompatible feedstock for the production of a double-layered vascular graft. The resulting graft exhibited nonlinear elasticity attributed to its flower-mimicking geometry. Natural polymers, including methacryloyl gelatin (GelMA) and sodium alginate, were employed for the vessel wall of the graft, demonstrating high biocompatibility and cell adhesion in vitro. By leveraging 3D printing techniques, this project efficiently achieved biomimetic mechanical properties while offering flexibility in implementing various geometries. These findings highlight the potential of individualized medicine and advanced treatment approaches for coronary artery disease.

732a Keynote Talk: Techno-economic and life cycle analyses of bioproducts across a large metabolic space: integrated analysis tools to advanced biomanufacturing sustainability

Pahola Thathiana Benavides, Bruno Klein

The development of biofuels and bioproducts, through the deployment of large-scale biorefineries, represents a multifaceted strategy to decarbonize the economy. Maximizing the potential of biofuels and bioproducts requires a comprehensive understanding of various factors, including technological feasibility, economic viability, and environmental sustainability. Under the Agile BioFoundry (ABF) consortium, a Department of Energy (DOE)-sponsored collaboration, we study several biobased pathways to produce advantaged products including fuel intermediates and bioproducts. The ABF integrates advanced computational tools for biological engineering, process design, data analysis, and sustainability modeling tools into a comprehensive and dynamic platform for biomanufacturing microbes and leveraging them to produce several products of industrial interest. In this work, we examine both the economic and environmental aspects of producing biobased molecules through fermentative processes and catalytic upgrading via techno-economic and life-cycle analyses. We discuss the key drivers of costs and greenhouse gas (GHG) emissions to identify the potential to achieve economic and environmental sustainability by meeting or exceeding commodity product benchmarks, thus prioritizing future R&D focus areas to help inform decision-making processes tied to the sustainability of new bio-based products.

732b Keynote Talk: Towards the Automated Optimal Control of Colloidal Self-Assembly

Xun Tang

Controlling colloidal self-assembly with external field-mediated non-equilibrium interferences is a viable approach to manufacturing desired structures, for advanced applications in energy harvest, drug delivery, and photonics devices. However, the scaleup and automation of this process is hindered by factors ranging from the identification of the system states to learning the system dynamics, and to the design of the optimal control policy. In this talk, I will focus on an electric field-mediated colloidal self-assembly system, to first talk about our previous efforts on how data can be used to learn the system dynamics in designing an optimal control policy to rapidly produce ordered colloidal crystals, and then discuss ongoing work on how we leverage advanced machine learning approaches to explore the automation of learning the optimal feedback control policy, especially on the state representation and policy design with a continuous action space.

732c High-Throughput Testing Approaches for Geopolymer: Discovering the Ideal Recipe

Vamsi Vikram Gande, Meenesh R Singh

Geopolymer concrete, derived from industrial mineral waste like coal ash and slag, offers a sustainable alternative to traditional Portland cement. However, the variability in feedstock composition poses a significant challenge in geopolymer production, requiring adjustments to process parameters such as alkali concentration and solid-to-liquid ratio for each feedstock. This study introduces a high-throughput strategy for optimizing the alkali activation process in forming geopolymer concrete from carbonated industrial mineral waste. Figure 1 outlines a three-step process: automated dispensing of carbonated residues and liquid alkali activator, reaction activation, and gel characterization. To achieve statistically relevant sample sizes, microtiter plates with 36 and 48 wells of approximately 1 cm diameter were designed and fabricated. A 3D-printed solid dispenser was used to accurately dispense fixed amounts (50 mg to 100 mg) of carbonated residue, while a robotic liquid dispenser (Dragonfly Discovery) added specific liquid solution volumes (0.01 ml to 0.1 ml) to the wells. The mixture was thoroughly mixed and compacted under high pressure at various temperatures. This study screened a range of reaction conditions including various alkali activators (NaOH, Na_2SiO_3, KOH, K_2SiO_3), solid-to-liquid weight ratios, pressure (500-3000 psi), and temperature (25°C to 70°C). The reaction between the liquid and powder components resulted in gel formation, which was characterized using FTIR spectrometer to analyze aluminosilicate network polymerization. The alkali activation process can be observed by shifting the main peak from 990 cm^{-1} to 920 cm^{-1}. This change is expected due to the decrease of tetrahedral Si sites, replaced by the tetrahedral Al during the formation of N-A-S-H gel. This provides the information about alkali activation rates. For a specific condition of L/S 150 ul/ 0.1 g, the change in peak intensities is shown in figure 2. This variation is captured for all the process conditions to determine the alkali activation rates.

Currently, we are trying to correlate the spectroscopy signatures and alkali activation rates to reactivity quantified by calorimetry, thermogravimetry, and, eventually mechanical performance. Once validated, this HT screening approach will be used to supply data for ML, which is expected to drastically accelerate mixture development for on-traditional concrete.

732d Image-Based Control of a Material Healing Process with Blender

Akkarakaran Leonard, Helen Durand

Industry 4.0 heralds the use of big data not only in the amounts of it generated, but also the different forms that can be captured, stored, and processed. With advances in computing capabilities, using image-based sensors and control algorithms on-site become a viable option for plant-wide applications. Image-based control is popular in the field of robotics, such as the control of unmanned aerial vehicles [1], and in chemical engineering applications, as in [2], where images are used to classify flow regimes in aerated tanks. A process where images can be extremely useful to determine states is Directed Self-Assembly (DSA), where individual particles arrange themselves into desired structures under an external control action. Image-based control of this process has been explored in the work of [3], where colloidal particles suspended in a liquid are assembled with an image-based, closed-loop control system. There is a need to be able to simulate such applications in a virtual environment, in order to benchmark and test a proposed system before deployment to the physical plant.

In this work, we present the use of Blender, a free and open source 3D modeling and animation software, as a test bed for image-based control applications. We have explored the use of Blender in similar applications, such as level control in a tank, disturbance handling due to weather phenomena, and the control of a nanorod under Brownian motion [4, 5]. Here we simulate and control the motion of a particle in a fluid medium in order to repair a layer present on the inner wall of a cylindrical pipe. Virtual cameras in Blender are used to capture images of the particle as it travels downstream to the repair site, and control actions are executed that cause the particle to move to the site in order to repair the layer. This simulation functions as a precursor to a self-healing corrosion inhibitor layer. This demonstrates the use of Blender to reconstruct the 3D location of the particle from images captured by separate cameras, and execute the control action using a Stochastic Model Predictive Controller (SMPC). The SMPC takes into the account the motion of the particle due to flow in the pipe, as well as Brownian motion deviations (modeled as random variables) due to impacts from particles in the medium.

References:

[1] Rafik Mebarki and Vincenzo Lippiello. “Image-based control for aerial manipulation”. In: Asian Journal of Control 16.3 (2014), pp. 646–656.
[2] Corinna Kr¨oger, Valentin Khaydarov, and Leon Urbas. “Data-driven, Image-based Flow Regime Classification for Stirred Aerated Tanks”. In: Computer Aided Chemical Engineering. Vol. 51. Elsevier, 2022, pp. 1363–1368.
[3] Xun Tang and Martha A Grover. “Control of microparticle assembly”. In: Annual Review of Control, Robotics, and Autonomous Systems 5 (2022), pp. 491–514.
[4] Henrique Oyama et al. “Development of directed randomization for discussing a minimal security architecture”. In: Digital Chemical Engineering 6 (2023), p. 100065.
[5] Akkarakaran Francis Leonard et al. “Virtual Test Beds for Image-Based Control Simulations Using Blender”. In: Processes 12.2 (2024), p. 279

732e Synergistic Integration of Reinforcement Learning with Conventional Process Control

Daniel Beahr, Debangsu Bhattacharyya

Reinforcement learning (RL) is a machine learning technique with high potential for application to autonomous systems. The flexibility and adaptability of RL algorithms make them an attractive alternative for control applications when compared to standard conventional control approaches. The model-free nature of RL eliminates the necessity for system identification. Furthermore, the more recent developments in the field of RL allow algorithms to capture complex continuous dynamic systems [1], [2], whereas previously, the simpler basis functions were often constrained by the dimensionality of systems. In addition, the adaptability of RL allows it to update the control policy over time, e.g., for capturing slowly changing dynamics which may necessitate a new system identification or updated tuning for conventional control methods. However, RL has its own drawbacks and shortcomings that have limited its implementation for control applications. Because of the black-box nature of RL, it is difficult to predict and analyze the behavior of RL algorithms. This is unacceptable for many performance-critical process control tasks. In addition, the significant exploration requirements for training RL make online implementation of the algorithm difficult. An RL with incomplete or insufficient training poses the risk of significant degradation of control performance. Because of these reasons, the maximization of the expected return while ensuring reasonable system performance without degradation has been a focus of published literature [3]. When implementation of a naïve RL agent to a system is unacceptable, it is often desired to couple it with another form of conventional process control (CPC) [4].

In this work, we propose a control structure by augmenting existing CPC methods with an RL agent implemented in parallel. Due to the generally slow learning rates and high exploration requirements of RL, it is desired to have the existing conventional process controller (e.g., PID, MPC, etc.) continue to compute its own generated control action to enhance the learning rate of the RL agent [5]. A weighted sum of the control actions of the RL and CPC is derived, and subsequently applied to the plant [6]; the resultant states and actions are then used to supplement the RL agent's learning. The proposed algorithm avoids direct action of the naive RL agent that may not result in acceptable performance and may even be unsafe under worst case scenarios. Algorithms are developed for an adaptive weighting function based on a measure of instantaneous and historical performance. Performance of both the RL and CPC methods are assessed using a moving horizon that decays with time, valuing more recent actions as more relevant than older actions. In addition, short-term trends are derived to allow for rapid transitions. In this way, the RL agent can take over control as and when its control performance exceeds that of the CPC method. If the RL's performance begins to deteriorate, the conventional control method would again assume full control before significant degradation of performance.

The algorithm is demonstrated on a dynamic process model of a solid oxide fuel cell (SOFC) plant for H_2 and power production [7]. The RL algorithm used is twin-delayed deep deterministic policy gradient (TD3) for temperature regulation at the outlet of the SOC stack. The TD3 approach offers a notable advantage in its capability to handle continuous action spaces and allows for direct one-to-one comparison with control actions from the conventional control method. For temperature regulation of the SOC system, the CPC is a series of PID controllers arranged in cascade loops. Because of the complex dynamics associated with SOC mode-switching operation between hydrogen and power production, the performance of the PID controllers can be poor whereas the actor-critic structure of the RL algorithm seeks to facilitate accurate capturing of the nonlinear dynamics. For this case study, the RL is proposed to augment, and eventually phase out, the cascaded PID loops. The episodic learning used for the RL-CPC arrangement is a series of hydrogen production set-point changes. Mode switching from maximum hydrogen production to maximum power production and back to maximum hydrogen production is considered as well. While learning is episodic in nature, the states are continuous across episodes, creating a consistent measure of performance improvement.

Specific contributions of this work include:

- An algorithm is developed for the parallel implementation of RL alongside conventional process control, allowing for transition of control from CPC to RL based on current and past performance. By leveraging short-term and long-term projections of control performance, this algorithm facilitates effective switching without degrading control.
- Online training and implementation of a direct RL algorithm for process control on systems of complex nonlinear continuous dynamics is demonstrated. It is shown that degradation of RL performance is limited throughout training to the level expected of in-place conventional control.
- It is observed that the RL-CPC algorithm can learn from and surpass a sub-optimal policy demonstrated by the conventional form of control that is in-place, eventually arriving at a superior policy than that of the conventional method.
- It is observed that the rate at which an RL-CPC arrives at an optimal policy surpasses traditional online RL methods with limited performance degradation.
- It is demonstrated that in cases where the RL encounters an unknown operational condition leading to degraded control performance, the control system will revert back to the conventional control that is in place, thereby limiting potential error and/or poor performance.

[1] T. P. Lillicrap *et al.*, "Continuous control with deep reinforcement learning," *4th Int. Conf. Learn. Represent. ICLR 2016 - Conf. Track Proc.*, Sep. 2016, [Online]. Available: http://arxiv.org/abs/1509.02971

[2] S. Fujimoto, H. Van Hoof, and D. Meger, "Addressing Function Approximation Error in Actor-Critic Methods," in *35th International Conference on Machine Learning, ICML 2018*, Feb. 2018, pp. 2587–2601. [Online]. Available:

http://arxiv.org/abs/1802.09477

[3] J. García and F. Fernández, "A comprehensive survey on safe reinforcement learning," *J. Mach. Learn. Res.*, vol. 16, pp. 1437–1480, 2015.

[4] O. Dogru *et al.*, "Reinforcement Learning in Process Industries : Review and Perspective," *2 IEEE/CAA J. Autom. Sin.*, vol. 11, no. 2, pp. 1–19, 2024, doi: 10.1109/JAS.2024.124227.

[5] J. A. Clouse, "On Integrating Apprentice Learning and Reinforcement Learning," University of Massachusetts, 1996.

[6] M. T. Rosenstein and A. G. Barto, "Reinforcement learning with supervision by a stable controller," *Proc. Am. Control Conf.*, vol. 5, pp. 4517–4522, 2004, doi: 10.1109/ACC.2004.182663.

[7] D. A. Allan *et al.*, "NMPC for Setpoint Tracking Operation of a Solid Oxide Electrolysis Cell System," *Found. Comput. Aided Process Oper. / Chem. Process Control (FOCAPO/CPC 2023)*, pp. 1–6, 2023, [Online]. Available: https://www.netl.doe.gov/projects/files/NMPCforSetpointTrackingOperationofaSolidOxideElectrolysisCellSystem_paper_011023.pdf

732f A Reinforcement Learning Framework for the Simultaneous Generation, Design and Control of Chemical Process Flowsheets

Simone Reynoso-Donzelli, Luis Ricardez-Sandoval

The integration and optimization of chemical process design and control systems hold a major significance in the field of engineering since it has the potential to improve the economic and sustainable operation of chemical process systems. The inherent complexities and challenges that arise when performing integration of design and control has prompted the search for more streamlined solutions. Traditionally, the sequential approach has been used to tackle optimal process design and control systems. In this framework, the steady-state problem aiming to determine equipment specifications and operations is addressed first followed by a controllability analysis aimed to assess the dynamic feasibility of the system in closed-loop. Despite its apparent effectiveness, this approach may not yield an optimal or even feasible transient operation since process control decisions are not considered at the design stage. Hence, there is a need for an integrated approach to design and control, ensuring the creation of a feasible plant flowsheets that meet design objectives consistently across both steady-state and transient phases of the plant in the presence of external disturbances [1]. A key challenge in chemical engineering applications involves the selection of integer process design decisions, e.g., number of trays in a distillation column (or the number of reactors) needed to achieve a particular product purity. Coupling integer design decisions with time-dependent process variables (e.g., systems' states) gives rise to a complex problem, known as mixed-integer dynamic optimization (MIDO). Significant progress has been made in this field, showcasing promising outcomes [1] – [3]. Despite these efforts, the current applications are limited to a relatively small number of processing units due to the high computational costs associated with the solution of integrated problems. Amid the rise of Machine Learning techniques in process systems engineering, numerous challenges are currently being addressed through model-free optimization approaches, prominently utilizing Reinforcement Learning (RL) techniques. This trend extends to issues involving the integration of process design and control. Recognizing the compelling outcomes presented in [4] – [5], this study endeavors to delve into the application of RL techniques for solving integrated design and control problems that involve integer (process design/flowsheet) decisions.

This study introduces a framework capable to simultaneously generate, design and control chemical process flowsheets using RL. A Proximal Policy Optimizer (PPO) agent is used to interact with the environment. The agent's primary goal is to simultaneously generate, design and control a chemical process flowsheet that can optimize a user-defined objective function, which also considers the process dynamic variability in closed-loop, e.g., disturbance rejection tracking errors, while adhering to the process and equipment operation and design constraints. Objectives and constraints are enforced using a reward shaping strategy. The agent interacts with an environment composed of different neural networks (NNs), which are identified *a priori* and aim to approximate the dynamic models that resemble the closed-loop operation of the unit operations included the integrated problem. The nature of the PPO agent depends on the action space of the integrated problem, i.e., if discrete, continuous or hybrid decisions are considered. Hence, multiple unit operations of a different kind can be considered by the PPO agent. For instance, if the problem involves more than one unit operation, a hybrid PPO agent that is capable of making discrete decisions (e.g., type of units in a flowsheet) and continuous decisions (e.g., unit operation's set-point) simultaneously can be designed. In this approach, the agent is tasked to interacting with the environment, i.e., adding unit operations to the flowsheet, until achieving the specified design and control goals. The dataset used to train the NNs is generated through a systematic process. Initially, a Latin hypercube is generated, encompassing all the model inputs, e.g., inlet stream variables, design variables, and control variables. Subsequently, every combination within this preliminary database is assessed within the dynamic process model to gather the corresponding outlet parameters and variables, e.g., species concentration and temperature of the outlet streams, along with a control tracking parameter and constraint tracking parameters. Following this evaluation, input/output data gathered from these simulations is tailored to fit a set of NNs (i.e., surrogate models) that are then integrated into the environment. A primary challenge in this methodology involved converting a transient profile into a singular explanatory value usable by the RL agent, given that all output variables in the dynamic process models are time-dependent. To address this issue, performance metrics borrowed from control theory were adopted in this work. The control tracking performance was assessed through the Integral of the Squared Error (ISE). The constraint tracking parameter was also assessed using the ISE; however, when translated into sub-rewards terms, it was accounted for as a penalty if it fell outside a pre-specified threshold. In cases where the interaction between the agent and the environment concludes with even minor violations in the constraint tracking parameter (resembling a constraint programming approach), it may result in suboptimal learning rates for the agent. Hence, this strategy was selected to encourage exploration by the agent and promote constraint satisfaction. The integration of these parameters showed an effective way to capture tracking errors and constraint violations in closed-loop, facilitating its utilization by the RL agent.

The RL methodology was applied to a case study aimed to perform simultaneous design and control of a process flowsheet for a first-order reaction subject to a step disturbance in the main inlet stream. The only unit operation available to the RL agent was a set of continuously stirred tank reactor (CSTR), whose thermodynamic parameters and dynamic model were adapted from the literature [6]. This case study was implemented on a PC with Intel® CoreTM i7-3770 CPU @ 3.40 GHz and 32 GB of RAM, using Python as the programming language. The main libraries used were torch for the development of the neural networks and gym for creating the RL environment. The objective of the agent was to simultaneously design and control the number of reactors required to achieve a 95% conversion of reactant, smoothly reject the inlet flow disturbance, and ensure that the reaction temperature remained below 400 K to prevent a runaway reaction within any of the CSTRs that may be included in the process flowsheet. Each CSTR considered in the flowsheet included a PI-controller that regulated the outlet concentration by adjusting the jacket's temperature. The dataset used to train the NNs embedded in the environment was generated using a Latin hypercube of 60,000 points on all the inlet variables for the reactor's model, i.e., inlet concentration and temperature, set-point concentration, volume, and the controllers' proportional gain and time integral. For this case study, the part that demanded the most computational resources was the identification of the surrogate NNs models, i.e., the simulation of the combinations included within the Latin hypercube design space. Once the dataset was completed with simulations of the dynamic CSTR, the output concentration and temperature, ISE, and temperature constraint tracking were adjusted for 3 regression NNs and one classification NN, respectively. The classification NN determined whether the operating temperature of the CSTR was within the feasible limits. The resulting NNs architecture was then embedded within the RL's environment. The PPO agent was trained for 100,000 steps over the course of one hour. The flowsheet obtained by the agent consisted of two reactors, which successfully dampened the inlet low disturbance and achieved a 95% conversion of the reactant in the second reactor. Outcomes achieved by the RL agent were compared against a model-based optimization methodology. Although the resolution process for the proposed framework may take longer, this methodology may be attractive to consider more complex scenarios involving multiple units of different kinds modelled using rigorous thermodynamic correlations and dynamic conservation balance equations. In such instances, approximating these models through NNs and subsequently optimizing them using the proposed RL methodology presents a highly attractive advantage over model-based optimization methods. This approach offers flexibility and adaptability in tackling intricate problems where traditional modeling approaches may fall short, showcasing the potential of RL in effectively addressing the simultaneous generation, design, and control of chemical flowsheets.

References

[1] Ricardez-Sandoval, L. A., Douglas, P. L., & Budman, H. M. (2011). A methodology for the simultaneous design and control of large-scale systems under process parameter uncertainty. Computers & Chemical Engineering, 35(2), 307–318. https://doi.org/10.1016/j.compchemeng.2010.05.010

[2] Koller, R. W., Ricardez-Sandoval, L. A., & Biegler, L. T. (2018). Stochastic back-off algorithm for simultaneous design, control, and scheduling of multiproduct systems under uncertainty. AIChE Journal, 64(7), 2379–2389. https://doi.org/10.1002/AIC.16092

[3] Rafiei, M., & Ricardez-Sandoval, L. A. (2020c). Integration of design and control for industrial scale applications under uncertainty: a trust region approach. Computers & Chemical Engineering, 141, 107006. https://doi.org/10.1016/J.COMPCHEMENG.2020.107006

[4] Sachio, S., del-Rio Chanona, A. E., & Petsagkourakis, P. (2021). Simultaneous Process Design and Control Optimization using Reinforcement Learning. IFAC-PapersOnLine, 54(3), 510–515. https://doi.org/10.1016/J.IFACOL.2021.08.293

[5] Mendiola-Rodriguez, T. A., & Ricardez-Sandoval, L. A. (2022). Robust control for anaerobic digestion systems of Tequila vinasses under uncertainty: A Deep Deterministic Policy Gradient Algorithm. Digital Chemical Engineering, 3, 100023. https://doi.org/10.1016/J.DCHE.2022.100023

[6] Seborg, D. E., T. F. Edgar, and D. A. Mellichamp, Process Dynamics and Control, 2nd Edition, Wiley, 2004, pp. 34–36 and 94–95.

664a Translating Process Flow Diagrams to Piping and Instrumentation Diagrams (PandIDs) with Artificial Intelligence

Dominik Goldstein, Lukas Schulze Balhorn, Edwin Hirtreiter, Artur M. Schweidtmann

Piping and instrumentation diagrams (P&IDs) are important depictions of chemical and biochemical processes containing schematics of a process' equipment, piping networks, and control structures. Developing P&IDs from process flow diagrams (PFDs) is a crucial step in process development, but typically involves a tedious, manual workflow.

We propose a methodology based on artificial intelligence for the prediction of control structures of P&IDs from PFDs, drawing inspiration from transformer-based models used in translating human language. This method treats the forecasting of control structures as a translation problem, transforming PFDs into P&IDs. Both flowsheets are represented as strings using the SFILES 2.0 notation. This makes them compatible with existing transformer-based translation models. Initially, our model undergoes pre-training on synthetic P&IDs to learn the syntax. Subsequently, it undergoes a finetuning step through transfer learning on industrial P&IDs.

References

[1] Hirtreiter, E., Schulze Balhorn, L., & Schweidtmann, A. M. (2024). Toward automatic generation of control structures for process flow diagrams with large language models. AIChE Journal, 70(1), e18259.

664b Developing a Framework for Enhanced Troubleshooting with Large Language Models

Jihan Abou Halloun, Anh Tran, Helen Durand

Large language models (LLM's) have received significant attention in recent months with the interest in the capabilities of ChatGPT. Large language models have been of interest to companies in a variety of applications, including in developing chatbots. Another more rigorous need for large language models is in the notion of "enhanced troubleshooting," by which we mean the ability of an individual at a company to query a centralized machine to find out why a certain failure scenario is occurring with, for example, equipment or materials. The concept behind enhanced troubleshooting is that a good deal of information is lost in industry as different individuals across different sites are faced with the same failure scenarios. Looking up why a particular problem may be occurring may require searching through a number of company databases that may not communicate with one another, or calling the right person at the company. It is also possible that an individual may not know who to contact, and consequently may end up solving a problem that has already been solved by another individual at the company all over again. These types of scenarios waste company resources and time.

LLM's offer a potential direction for alleviating these issues. Specifically, they can have access to a centralized databased of company information from which they draw to provide answers to user queries in an automated workflow. This workflow may include accessing the database of information to pull out relevant documents, checking for relevancy, and then using the LLM to synthesize a response to the user's query from the relevant documents. However, currently, this workflow would suffer from key challenges with LLM's today, in particular that there is not a clear procedure, when engineering decision-making is in view, for assessing the relevancy of documents to an engineering task. Therefore, in this work, we discuss the enhanced troubleshooting workflow and the unique characteristics that it brings, such as the need to avoid fine-tuning, which limit the options for relevancy testing. We then discuss a number of different options for attempting to improve relevancy testing that involve claim simplification and evidence retrieval, pointwise [1] and listwise rerankers [2,3], sparse, dense, and hybrid searching [4] for the most relevant documents, and question-and-answer protocols [5] meant to bring out similarities and differences between information retrieved from the database and the original user's question to aid with relevancy testing. We compare the results of using these different strategies for a variety of queries to evaluate the significance of these various potential components of the enhanced troubleshooting workflow to the overall goals of the method. Through this, we develop insights into how to create enhanced troubleshooting systems for engineering problems.

References:

[1] Ma, Xueguang, Liang Wang, Nan Yang, Furu Wei, and Jimmy Lin. 'Fine-tuning llama for multi-stage text retrieval.' *arXiv preprint arXiv:2310.08319* (2023).

[2] Ma, Xueguang, Xinyu Zhang, Ronak Pradeep, and Jimmy Lin. 'Zero-shot listwise document reranking with a large language model.' *arXiv preprint arXiv:2305.02156* (2023).

[3] Pradeep, Ronak, Sahel Sharifymoghaddam, and Jimmy Lin. 'RankZephyr: Effective and Robust Zero-Shot Listwise Reranking is a Breeze!.' *arXiv preprint arXiv:2312.02724* (2023).

[4] Bhagdev, Ravish, Sam Chapman, Fabio Ciravegna, Vitaveska Lanfranchi, and Daniela Petrelli.

'Hybrid search: Effectively combining keywords and semantic searches.' In *The Semantic Web: Research and Applications: 5th European Semantic Web Conference, ESWC 2008, Tenerife, Canary Islands, Spain, June 1-5, 2008 Proceedings 5*, pp. 554-568. Springer Berlin Heidelberg, 2008.

[5] Yang, Jing, Didier Vega-Oliveros, Taís Seibt, and Anderson Rocha. 'Explainable fact-checking through question answering.' In *ICASSP 2022-2022 IEEE International Conference on Acoustics, Speech and Signal Processing (ICASSP)*, pp. 8952-8956. IEEE, 2022.

664c Accelerated Process Optimization for a Circular Economy Using Full Flowsheet Neural Network Models

Ville Kaerkaes, Tim Kircher, Maximilian Hungsberg, Martin Votsmeier, Bastian Etzold

With most of chemical industry having been designed around a linear value chain, the transition to a circular economy poses significant challenges: Recycling waste in feed streams results in fluctuations of feed compositions. The optimization and testing of processes with regard to their tolerance of these fluctuations are therefore of great importance to ensure resilience in the changing value chains.[1] Still, optimization by high-throughput flowsheet simulation is held back by significant computational cost. Neural network surrogate models have the potential to drastically reduce this cost.[2] Here, an automated high throughput flowsheet simulation was realized for a process describing plastic waste gasification and subsequent methanol synthesis. This simulation was used to generate data for training and testing neural networks as surrogate models for the complete flowsheet as a function of both feed compositions and process parameters. Then, the surrogate model is used for optimization of process parameters at a significantly reduced computational cost compared to the flowsheet simulation.

A candidate process for methanol production by plasma torch gasification of plastic waste was simulated using ASPEN Plus® V12.1 with varying feedstock compositions and process parameters. The flowsheet was constructed to be simple and rigid, with parameters typically set with design specifications chosen as parameters to be varied. Feedstock compositions were taken from an established elemental distribution of plastic waste[3] and the selected process parameters were randomly varied within selected bounds. The resulting methanol mass flows and gasifier heat duties were used as training targets in the dataset. A total of 20,000 data points were used to train and test the model. The simulations were run automatically using an Python routine with robust error handling. The dataset was used to train a feedforward neural network (FNN) implemented in the scientific computing Python library JAX. Gradient-based constrained optimization is utilized to maximize profits as a function of process parameters. The model was applied in a case study using the average EU plastic waste composition, aiming to demonstrate surrogate-based multi-objective optimization.

The trained FNN model achieves average relative errors of 2.0 % and 1.4 % for the methanol product stream and gasifier heat duty respectively, showing the ability to generalize based on the training data. A parity plot for the testing dataset is shown in Figure 1. In a next step, the surrogate models are used to optimize process parameters by gradient based optimization to maximize the process profit. The JAX library enables using automatic differentiation to calculate analytical gradients with little computational overhead. For the case study, a single optimization takes an average of 0.17 seconds on a consumer grade CPU while calling the neural network surrogate an average of 128 times. This means that the FNN model is at least three orders of magnitude faster than the flowsheet simulation. This low simulation cost enables high-throughput experimentation, allowing for optimization across a wide variety of scenarios, for example for a varying electricity price as shown in Figure 2.

Neural networks show great promise as surrogate models in process engineering due to their ability to handle multidimensional problems. The shown model achieves average relative errors of 2.0 % and 1.4 % with a dataset of 20,000 datapoints on a complex problem, indicating scalability for more extensive flowsheets. We demonstrate the ability to optimize for varying conditions, enabling increased process tolerance in fluctuating environments. The results of the case study show the potential of the model for process optimization, as it is at least three orders of magnitude faster than the original flowsheet simulation. The chemical industry stands to benefit greatly from the further development of AI surrogate models, which have the potential to significantly reduce the computational cost of simulations and digital twins while maintaining accuracy. This represents a major step towards Industry 4.0.

[1] A. Mitsos, N. Asprion, C. A. Floudas, M. Bortz, M. Baldea, D. Borvin, A. Caspari, P. Schäfer, *Computers & Chemical Engineering* **2018**, 113, 209.

[2] M. Gärtler, V. Khaydarov, B. Klöpper, L. Urbas, *Chemie Ingenieur Technik* **2021**, *93*, 2063.

[3] M. Hungsberg, *Distribution of European plastic waste composition*, Technical University of Darmstadt, **2020**.

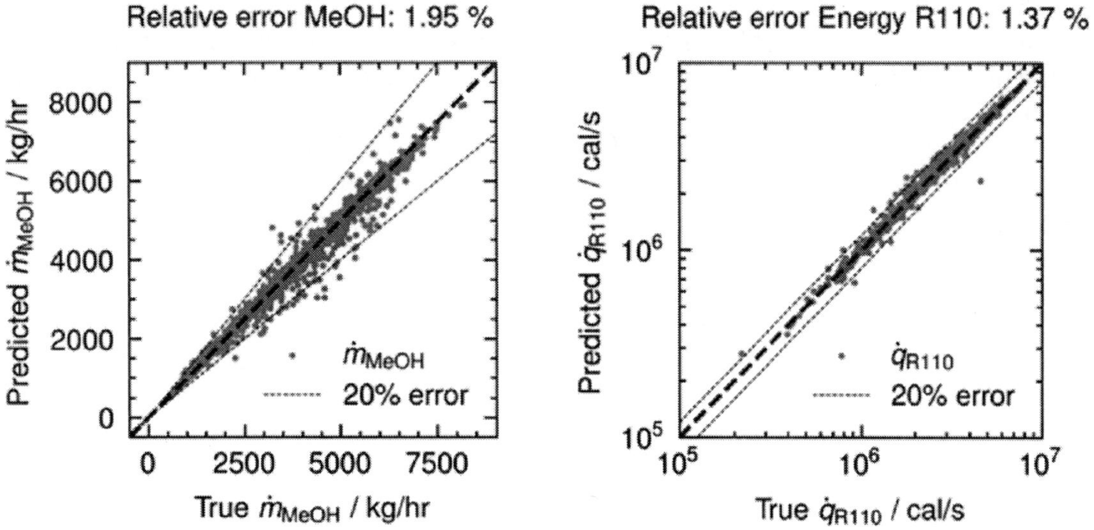

Figure 1: Parity plots of ground truth and neural network prediction on the test dataset for MeOH product massflow and gasifier heat duty.

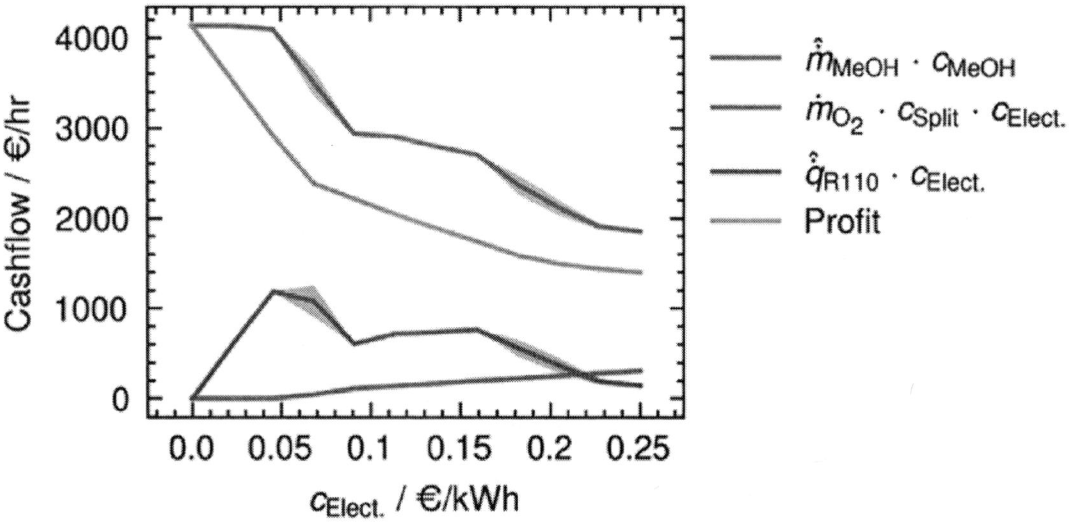

Figure 2: Process feasibility study for average EU plastic waste feed as a function of the electricity price derived by multi-objective optimization of the profit. Optimization of process parameters is performed over 25 different, randomly selected initial parameters and confidence intervals are shown.

664d Prompt Engineering-Directed in-Context Learning of an LLM with Integrated Automated Simulation for Generative Chemical Process Design

Donghyeon Lee, Dongil Shin

Chemical process design involves the optimal design of large-scale chemical processes among various candidates and requires considerable time and effort to navigate extensive search space constrained by cost-effectiveness, safety, environmental friendliness, and market conditions. Various AI-based methodologies have been proposed, such as finding the optimal process through reinforcement learning or outputting the best process through greedy search and simulations. However, these methodologies exhibit two critical drawbacks: low versatility, making it challenging to apply the methodologies to diverse processes, and low explainability, making it hard to explain how the optimal process was chosen.

This study proposes a methodology to overcome the low versatility and explainability by incorporating a Large Language Model (LLM), Generative Pre-trained Transformer (GPT), into the chemical process design process. To conduct process diagram input and analysis through a language model, we use the Detailed Simplified Flowsheet Input Line Entry System format that includes details of flows and equipment. Furthermore, we apply various prompt engineering techniques to direct the in-context learning of the LLM, integrated with automatically generated process simulations, to transform the general solver GPT into an intelligent agent specialized in chemical process design. Prompt chaining is applied to divide and solve the complex problem of chemical process design sequentially, breaking down the design and improvement process. Techniques such as retrieval augmented generation (RAG) for diverse processes application and In-Context Learning (ICL), automatic reasoning and tool-use (ART) for simulation integration, and directional stimulus prompting to focus on important information are applied to construct prompts specialized for chemical process design and to proceed with the enhanced design and improvement processes.

By applying the proposed system to various case studies, we conducted process design and improvement, demonstrating process improvements through various methods such as reactor addition and recycling. The improved results showed that diverse generic process improvements are possible. Moreover, the improved process was analyzed through the language model's inherent explanation capability and the additional integration with the simulator DWSIM, showing improved explainability. Ultimately, the study presents a new methodology for automating and supporting process design based on intelligent generation and validation of alternative processes guided by LLM-based design agents.

664e Geometry-Informed Deep Learning Surrogate Models for Flow Prediction and Optimisation of Flow Reactors

Nausheen Basha, Mosayeb Shams, Sibo Cheng, Rossella Arcucci, Omar Matar

Flow reactors, also known as continuous flow reactors, are used in chemical synthesis, pharmaceutical manufacturing or polymerisation. The shift from batch to continuous processes has led to the development of novel reactor configurations, such as 3D printed mesoscale reactors. These reactor designs, which are otherwise difficult to manufacture through other conventional techniques, have demonstrated high yields [1][2][3]. However, identifying optimal designs requires efficient computational methods to explore the vast design space effectively.

Computational Fluid Dynamics (CFD), CFD-led explorations have been extensively carried out for improving the design of reactors, where design improvements can be identified based on visualisations of the flows [4][5][6]. However, these simulations are expensive and are often based on human intuition to navigate the parameter space, which makes this approach ineffective at balancing the number of evaluations needed to identify truly optimal solutions. To address this, we have previously explored derivative-free Bayesian optimisation models, along with multi-fidelity optimisation techniques integrated with the CFD simulations. These approaches have shown promising results in uncovering crucial flow phenomena, such as Dean vortices, which enhance mixing at low Reynolds numbers [7][8][9]. With an aim to further reduce the costs of exploration, we propose to extend the multi-fidelity Bayesian optimisation by exploring fidelities encompassing full CFD as high-order models and reduced-order models through Geometry-Informed Graph Neural Networks (GNNs). The focus of this work is on developing and applying geometry-informed GNNs to efficiently predict flow dynamics across diverse reactor configurations and operating conditions.

GNNs have been successfully applied within chemical reactors to improve reaction yield prediction using pre-trained models [10]. They have also demonstrated the capability to learn fluid dynamics for various geometries, including airfoils, flow around cylinders [11], shapes generated through Bezier curves [12], and even wind turbine blades [13]. Our approach introduces a geometry-informed deep learning surrogate modeling framework that leverages GNNs to encode topological and geometric information, coupled with Convolutional Autoencoders for compressed flow representations. This framework enables us to replace costly CFD computations with efficient prediction mechanisms, significantly reducing computational overhead. Figure shows the GNN-Autoencoder-OpenFoam framework used in this study.

We pre-train the GNN model on a two-dimensional serpentine reactor with shape variants generated using sinusoidally-parameterised geometries. The geometric parameters varied include the amplitude ($p1 = 0.1$-0.5), frequency ($p2 = 3.0$-4.0), and horizontal offset ($p3 = 0$-$\pi/2$) of the sinusoidal curve. Structured meshes for each reactor geometry are created using the blockMesh utility in OpenFOAM v2012. The reactor geometry is defined by two sinusoidal curves representing the upper and lower walls, with smooth transitions at the inlet and outlet regions. Each geometry consists of approximately 4000 cells. We integrate OpenFOAM with Python using the PyFoam library.

We run transient CFD simulations using scalarTransportFoam to track the tracer concentration in the water medium. The flow is governed by the convection-diffusion equation, with convection being the dominant transport mechanism. The plug flow performance is evaluated through the residence time distribution (RTD), obtained from the tracer concentration at the outlet. More than 100 cases are generated using Latin-hypercube sampling. For each transient case, 140 arrays are generated to capture a time period of 7 seconds, with data recorded every 0.05 seconds.

The GNN architecture comprises graph convolution layers that operate on the reactor mesh graph, capturing spatial dependencies and encoding geometric information. The node features include x and y coordinates, tracer concentration, and velocity components. The graph convolution layers are followed by pooling layers to reduce spatial dimensions and extract high-level features. The compressed flow representations are then fed into fully connected layers to predict velocity and tracer concentration fields at future time steps. The model is trained using a combination of mean squared error loss for flow predictions and a reconstruction loss for the autoencoder.

Our integrated framework demonstrates good accuracy and efficiency in flow predictions. On a test set of 10 unseen reactor geometries, the GNN surrogate model achieves low mean squared errors for both tracer concentration and velocity predictions compared to ground truth CFD simulations. Additionally, the GNN surrogate model provides more than a 100-fold speedup in computational time compared to traditional CFD simulations, enabling real-time flow predictions and efficient design space exploration.

This geometry-informed deep learning surrogate modelling approach showcases the potential for accurate and efficient flow predictions, real-time simulations, and uncertainty quantification across complex reactor geometries and can be extended to three-dimensional geometries. Future work will focus on integrating this approach with the multi-fidelity optimisation framework, reporting on optimisation time, and validating the designs through 3D printing and experimental testing of the reactors.

Acknowledgements: This work is supported by the EPSRC PREMIERE (EP/T000414/1) Programme Grant.

References

[1] J. R. McDonough, J. Armett, R. Law, and A. P. Harvey, "Coil-in-Coil Reactor: Augmenting Plug Flow Performance by Combining Different Geometric Features Using 3D Printing," *Ind. Eng. Chem. Res.*, vol. 58, no. 47, 2019, doi: 10.1021/acs.iecr.9b04239.

[2] M. J. Harding *et al.*, "3D printing of PEEK reactors for flow chemistry and continuous chemical processing," *React. Chem. Eng.*, vol. 5, no. 4, 2020, doi: 10.1039/c9re00408d.

[3] M. C. Maier *et al.*, "3D Printed Reactors for Synthesis of Active Pharmaceutical Ingredients in Continuous Flow," *Org. Process Res. Dev.*, vol. 24, no. 10, 2020, doi: 10.1021/acs.oprd.0c00228.

[4] Z. Bao, F. Yang, Z. Wu, S. Nyallang Nyamsi, and Z. Zhang, "Optimal design of metal hydride reactors based on CFD-Taguchi combined method," *Energy Convers. Manag.*, 2013, doi: 10.1016/j.enconman.2012.07.027.

[5] J. Ding, X. Wang, X. F. Zhou, N. Q. Ren, and W. Q. Guo, "CFD optimization of continuous stirred-tank (CSTR) reactor for biohydrogen production," *Bioresour. Technol.*, 2010, doi: 10.1016/j.biortech.2010.03.146.

[6] S. Hapke, G. A. Luinstra, and K. M. Zentel, "Optimization of a 3D-printed tubular reactor for free radical polymerization by CFD," *J. Flow Chem.*, 2021, doi: 10.1007/s41981-021-00154-5.

[7] N. Basha, T. Savage, J. McDonough, E. A. del Rio Chanona, and O. K. Matar, "Discovery of mixing characteristics for enhancing coiled reactor performance through a Bayesian optimisation-CFD approach," *Chem. Eng. J.*, vol. 473, 2023, doi: 10.1016/j.cej.2023.145217.

[8] T. Savage, N. Basha, J. McDonough, O. K. Matar, and E. A. del Rio Chanona, "Multi-fidelity data-driven design and analysis of reactor and tube simulations," *Comput. Chem. Eng.*, vol. 179, 2023, doi: 10.1016/j.compchemeng.2023.108410.

[9] T. Savage, N. Basha, J. McDonough, O. K. Matar, and E. A. del Rio-Chanona, "Machine Learning-Assisted Discovery of Novel Reactor Designs via CFD-Coupled Multi-fidelity Bayesian Optimisation," 2023. [Online]. Available: https://doi.org/10.48550/arXiv.2308.08841.

[10] S. W. Li, L. C. Xu, C. Zhang, S. Q. Zhang, and X. Hong, "Reaction performance prediction with an extrapolative and interpretable graph model based on chemical knowledge," *Nat. Commun.*, vol. 14, no. 1, 2023, doi: 10.1038/s41467-023-39283-x.

[11] Z. Yang, Y. Dong, X. Deng, and L. Zhang, "AMGNET: multi-scale graph neural networks for flow field prediction," *Conn. Sci.*, vol. 34, no. 1, 2022, doi: 10.1080/09540091.2022.2131737.

[12] A. Gladstone, R. J., Rahmani, H., Suryakumar, V., Meidani, H., & Zareei, "Mesh-based GNN surrogates for time-independent PDEs," *Sci. Rep.*, vol. 14, 2024, [Online]. Available: https://doi.org/10.1038/s41598-024-53185-y.

[13] S. Li, M. Zhang, and M. D. Piggott, "End-to-end wind turbine wake modelling with deep graph representation learning," *Appl. Energy*, vol. 339, 2023, doi: 10.1016/j.apenergy.2023.120928.

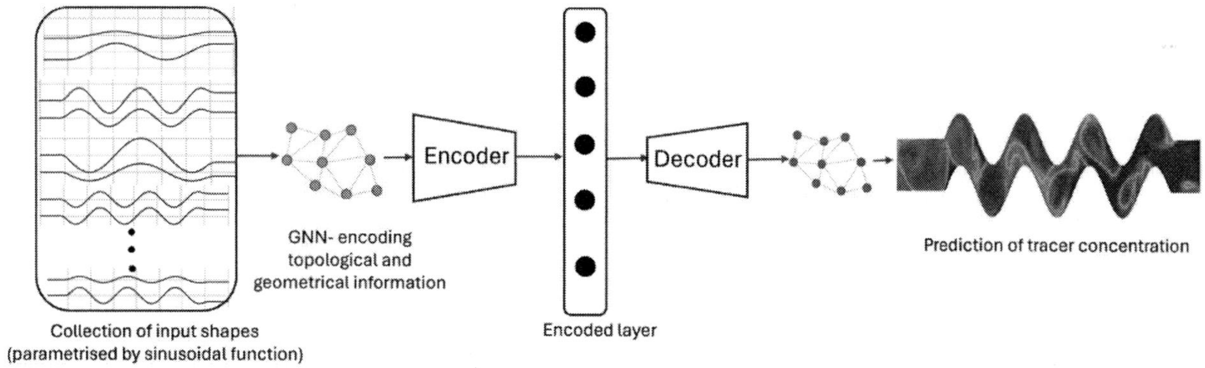

Figure 1: Illustration of the GNN-Autoencoder-OpenFOAM framework used in this study.

664f Dynamic Graph-Based Distributed Estimation and Control of Fast-Evolving Complex Process Networks

Amirmohammad Ebrahimi, Davood Babaei Pourkargar

Model Predictive Control (MPC) is widely utilized in chemical process manufacturing and energy systems for executing optimal decision-making and control actions, owing to its adaptability, inherent robustness, and capability to effectively manage complex multivariable systems subject to operational and safety constraints [1]. MPC formulates the control problem as a dynamic optimization problem that must be solved frequently at every sampling interval to determine the optimal set of manipulated inputs [2,3]. Therefore, the feasibility of using MPC hinges on its ability to solve the underlying dynamic optimization problem in real-time, subject to the process model and constraints. This restricts the MPC implementation in a centralized manner, particularly for large-scale integrated processes. Distributed Model Predictive Control (DMPC) presents a viable alternative to the centralized MPC approach, potentially speeding up optimization through parallel computing while preserving a desired degree of control performance. DMPC decomposes the optimal control problem into smaller subproblems solved by cooperative control agents.

Decomposing a system is the initial step in implementing a distributed control architecture, identifying the optimal number of subsystems and the distribution of inputs and outputs among them. Effective decompositions are those that reduce the required communication between control agents, thereby lowering computational demands. From the network theory perspective, this problem can be represented by identifying subsystems with minimal interconnections but strongly related variables [4]. Methods like input-output connectivity clustering have been suggested to create a system decomposition hierarchy [5]. Moreover, community detection approaches have been applied to the graph representation of the systems to identify highly interconnected subsystems by maximizing a modularity index, which measures the extent of edge concentration within communities compared to a hypothetical random graph [7-9]. Case studies were conducted on various process networks, including a benchmark reactor-separator, benzene alkylation with ethylene, and an amine gas treatment facility, exploring diverse optimization frameworks, communication strategies among local controllers, and model uncertainty considerations. These studies have demonstrated that community detection-based decompositions of unweighted graph representations can accelerate computations without substantially compromising performance compared to the centralized approach [10-13]. Our recent work has shown superior distributed estimation and control performance utilizing weighted graph-based system decomposition over unweighted [14,15], while the weighted graphs only captured variable interactions at a steady state.

This work focuses on developing an adaptive distributed estimation and control architecture using dynamic graph-based community detection, where the distributed structure is recursively modified at each sampling time. Modularity is employed as the primary metric for community detection. However, the NP-hardness of maximizing modularity makes resolving at each sampling time impractical. Our alternative approach involves pre-clustering and updating the subsystems as needed during the transition. The proposed decomposition approach limits the number of variables allowed to participate in the community update at each sampling time. The proposed dynamic graph-based system community detection approach is then employed to develop an integrated distributed moving horizon estimation (DMHE) and DMPC design to address the output tracking problem. The adaptive distributed estimation and control approach is implemented for a benzene alkylation process benchmark. The closed-loop performance illustrates that employing dynamic weighted graph models for community detection of the integrated process representation enhances distributed state estimation and output tracking, surpassing the traditional approaches based on unweighted and steady state-weighted decompositions.

References:

[1] Rawlings, J. B.; Mayne, D. Q.; Diehl, M. M. Model Predictive Control: Theory, Computation, and Design 2nd Edition; 2019.

[2] Mayne, D. Q. Model Predictive Control: Recent Developments and Future Promise. *Automatica*, 2014, 50 (12), 2967–2986.

[3] Qin, S. J.; Badgwell, T. A. A Survey of Industrial Model Predictive Control Technology. *Control Eng. Pract.* 2003, 11 (7), 733–764.

[4] Girvan, M.; Newman, M. E. J. Community Structure in Social and Biological Networks. *Proc. Natl. Acad. Sci. U.S.A.* 2002, 99 (12), 7821–7826.

[5] Heo, S.; Daoutidis, P. Control-Relevant Decomposition of Process Networks via Optimization-Based Hierarchical Clustering. *AIChE J.* 2016, 62 (9), 3177–3188.

[6] Jogwar, S. S.; Daoutidis, P. Community-Based Synthesis of Distributed Control Architectures for Integrated Process Networks. *Chem. Eng. Sci.* 2017, 172, 434–443.

[7] Leicht, E. A.; Newman, M. E. J. Community Structure in Directed Networks. *Phys. Rev. Lett.* 2008, 100 (11), 118703.

[8] Pourkargar, D. B.; Almansoori, A.; Daoutidis, P. Impact of Decomposition on Distributed Model Predictive Control: A Process Network Case Study. *Ind. Eng. Chem. Res.* 2017, 56 (34), 9606–9616.

[9] Tang, W.; Pourkargar, D. B.; Daoutidis, P. Relative Time-Averaged Gain Array (RTAGA) for Distributed Control-Oriented Network Decomposition. *AIChE J.* 2018, 64 (5), 1682–1690.

[10] Pourkargar, D. B.; Almansoori, A.; Daoutidis, P. Comprehensive Study of Decomposition Effects on Distributed Output Tracking of an Integrated Process over a Wide Operating Range. *Chem. Eng. Res. Des.* 2018, 134, 553–563.

[11] Moharir, M.; Pourkargar, D. B.; Almansoori, A.; Daoutidis, P. Distributed Model Predictive Control of an Amine Gas Sweetening Plant. *Ind. Eng. Chem. Res.* 2018, 57 (39), 13103–13115.

[12] Pourkargar, D. B.; Moharir, M.; Almansoori, A.; Daoutidis, P. Distributed Estimation and Nonlinear Model Predictive Control Using Community Detection. *Ind. Eng. Chem. Res.* 2019, 58 (30), 13495–13507.

[13] Moharir, M.; Pourkargar, D. B.; Almansoori, A.; Daoutidis, P. Graph Representation and Distributed Control of Diffusion-Convection-Reaction System Networks. *Chem. Eng. Sci.* 2019, 204, 128–139.

[14] Ebrahimi, A.; Pourkargar, D. B. Distributed Model Predictive Control of Integrated Process Networks Based on an Adaptive Community Detection Approach. *In Proceedings of the American Control Conference, 2024,* in press.

[15] Ebrahimi, A.; Pourkargar, D. B. Distributed Estimation and Control of Process Networks using

Adaptive Community Detection. *In Proceedings of the IFAC Symposium on Advanced Control of Chemical Processes, 2024,* in press.

664g Active Learning and Symbolic Regression for Interpretable Modeling of Laser Power Bed Fusion (L-PBF) Additive Manufacturing Processes

Alexander Summers, Q. Peter He

Laser power bed fusion (L-PBF) is an emerging additive manufacturing (AM) technology in which metal parts are built layer-by-layer through laser sintering of metal powder from computer-aided design models [1], [2]. AM technology has progressed beyond prototyping in the past few decades and is now used to reliably produce parts for various industrial sectors. Additive manufacturing has the benefit over subtractive and formative manufacturing methods in that it can produce parts much more quickly while wasting little material. Recently, pulsed wave (PW) laser emission, as compared to the traditional continuous wave (CW) emission, has been proposed to refine the microstructure and improve the density and porosity of the manufactured parts. At the same time, PW emission brings a much larger parameter space, which can be difficult to traverse without proper techniques [3]. In the L-PBF process, there are generally three melting modes: (1) the conduction mode, where heating is dominated by the conduction of heat through the metal; (2) the keyhole mode, where heating is dominated by radiative heat transfer; and (3) the transition mode, where both phenomena compete for dominance. Previous work has shown that the transition mode is where the highest density is found/manufactured [1], [4]. In this work, we develop a process map for Ti-6V-4Al to identify where the transition mode region exists in the process parameter space. We do so by performing single scan tracks (SSTs) on titanium substrates and measuring the melt pool aspect ratios from cross-sectional imaging.

Working towards this goal, we use active learning (AL) through an iterative design of experiments (DOE) approach to maximize information gain from the process. Through this iterative DOE approach, we hope to gain as much information as possible from the parameter space in relatively few experiments. We combine this experimental case with other simulated cases to investigate whether specific AL strategies can be applied more generally. Among the AL strategies we implement are Gaussian process uncertainty sampling and leave-one-out sensitivity sampling, which we compare with random sampling and distance-based sampling. Both of these AL strategies attempt to sample from regions of the parameter space in which the output is either not well characterized or is highly sensitive to the process parameters [5], [6]. Finally, we use symbolic regression to obtain physically-interpretable models from the acquired data, so that we can gain insight into the physical phenomena occurring in the process. In this work, we show that we can use AL and symbolic regression to derive robust, physically-interpretable models in relatively few experiments. We show that we can extend this approach to other application areas to efficiently discover models that extract physical meaning from the systems of interest.

[1] S. Patel and M. Vlasea, "Melting modes in laser powder bed fusion," *Materialia (Oxf)*, vol. 9, Mar. 2020, doi: 10.1016/j.mtla.2020.100591.

[2] S. S. Razvi, S. Feng, A. Narayanan, Y.-T. T. Lee, and P. Witherell, "A Review of Machine Learning Applications in Additive Manufacturing," in *International design engineering technical conferences and computers and information in engineering conference*, Aug. 2019.

[3] E. Toyserkani, D. Sarker, O. O. Ibhadode, F. Liravi, P. Russo, and K. Taherkhani, "Basics of Metal Additive Manufacturing," in *Metal Additive Manufacturing*, Wiley, 2021, pp. 31–87. doi: 10.1002/9781119210801.fmatter.

[4] S. Patel and M. L. Vlasea, "Melting Mode Thresholds in Laser Powder Bed Fusion and their Application Towards Process Parameter Development," 2019.

[5] T. Lookman, P. V. Balachandran, D. Xue, and R. Yuan, "Active learning in materials science with emphasis on adaptive sampling using uncertainties for targeted design," *npj Computational Materials*, vol. 5, no. 1. Nature Publishing Group, Dec. 01, 2019. doi: 10.1038/s41524-019-0153-8.

[6] Q. Zhou *et al.*, "An active learning radial basis function modeling method based on self-organization maps for simulation-based design problems," *Knowl Based Syst*, vol. 131, pp. 10–27, Sep. 2017, doi: 10.1016/j.knosys.2017.05.025.

702a Keynote 1 - Resilient Multi-Agent Estimation and Control of Complex Process Networks

Davood Babaei Pourkargar

Cyber-physical systems, representing complex integrations between physical and digital elements, are particularly vulnerable to cyber incidents that can cause significant operational disruptions and safety hazards [1,2]. The growing complexity and nonlinearity of these systems, combined with the necessity for extensive communication, often render centralized predictive models impractical for incident detection, recovery, and control [3]. Traditional centralized approaches for process incident detection, parameter estimation, and control demand significant computational resources. Conversely, decentralized methods, where independent agents manage each subsystem, face challenges such as communication overhead, frequent synchronization, and complexity, especially in highly integrated systems [4-7].

This study proposes a novel adaptive multi-agent architecture for distributed incident detection, state recovery, and regulation. By leveraging spectral community detection, the method identifies highly interactive state variables within the system [8]. It recursively decomposes the system into a predetermined number of subsystems in a single step, differing from traditional hierarchical methods that sequentially divide subsystems to achieve maximum modularity. For incident detection, community-based long short-term memory (LSTM) models are developed using data from each community to predict the system's reference dynamics. Each LSTM model learns the normal operations of its community while considering integrated dynamics, thereby maintaining high detection accuracy while mitigating issues related to data collection, communication, and computational constraints.

A combined spectral community detection-based distributed moving horizon estimation (DMHE) and distributed model predictive control (DMPC) architecture is then utilized to further enhance the resilience of integrated process systems in the presence of cyber incidents. The DMHE estimates unmeasured state variables and recovers the falsified variables required for output feedback regulations. The proposed architecture is evaluated through simulations of an integrated benzene alkylation benchmark under various process incident scenarios. These evaluations demonstrate the proposed multi-agent architecture's closed-loop performance and computational advantages in detecting and mitigating cyber incidents, particularly those targeting temperature measurement sensors.

References:

[1] Zhe Wu, Scarlett Chen, David Rincon, and Panagiotis Christofide, Post cyber-attack state reconstruction for nonlinear processes using machine learning, Chemical Engineering Research and Design, 159, 248-261, 2020

[2] Henrique Oyama and Helen Durand. Integrated cyberattack detection and resilient control strategies using Lyapunov-based economic model predictive control. AIChE Journal, 66(12), e17084, 2020.

[3] Brayden Sundberg and Davood B. Pourkargar. Cyberattack awareness and resiliency of integrated moving horizon estimation and model predictive control of complex process networks. In Proceedings of the American Control Conference (ACC), 3815–3820, 2023.

[4] Scarlett Chen, Zhe Wu, and Panagiotis D. Christofides. Cyber-security of centralized, decentralized, and distributed control-detector architectures for nonlinear processes. Chemical Engineering Research and Design, 165, 25-39, 2021.

[5] Dana Alsagheer, Lei Xu, and Weidong Shi. Decentralized machine learning governance: Overview, opportunities, and challenges. IEEE Access, 2023.

[6] Yuwei Sun, Hideya Ochiai, and Hiroshi Esaki. Decentralized deep learning for multi-access edge computing: A survey on communication efficiency and trustworthiness. IEEE Transactions on Artificial Intelligence, 3(6):963–972, 2021.

[7] Prashant Mhaskar, Adiwinata Gani, Nael H. El-Farra, Charles McFall, Panagiotis D. Christofides, and James F Davis. Integrated fault-detection and fault-tolerant control of process systems. AIChE Journal, 52(6):2129–2148, 2006.

[8] AmirMohammad Ebrahimi and Davood B. Pourkargar. Distributed estimation and control of process networks using adaptive community detection. In Proceedings of 12th IFAC International Symposium on Advanced Control of Chemical Processes, 2024.

702b Keynote 2 - Approaches for Active Detection of False Data Injection Attacks Via Process Control System Design

Shilpa Narasimhan, Nael El-Farra, Matthew Ellis

Modern day process control systems (PCSs) may integrate new technologies, including Internet of Things (IoT) and cloud computing into chemical manufacturing facilities to enhance the productivity [1]. Technologies such as IoT may allow malicious actors to side-step information technology-based protections and alter data over PCS communication channels to compromise manufacturing operations using false data injection (FDI) cyberattacks [2]. Design of controller-based approaches for cyberattack detection, identification, and mitigation for the fortification of PCS cybersecurity have received interest in the literature [3], [4].

Cyberattack detection approaches are an important component of PCS cybersecurity fortification and detect an attack based on anomalies in data over PCS communication channels [5]. Broadly, detection approaches may be categorized as: passive detection schemes which monitor for attacks without an external intervention (e.g., [6]) and active detection methods which utilize an external intervention to detect an attack (e.g., [7]). The design of a detection approach may consider the class of FDI attack to be detected (see [10] for an elaborate taxonomy of FDI attacks). Passive detection schemes have been proposed for the detection of multiplicative [9] and additive FDI attacks [7], under which data over the compromised PCS communication channel is altered by the multiplication or addition of a factor, respectively. A passive detection method may fail to detect FDI attacks under which the behavior of the altered PCS data is indistinguishable from its attack-free behavior (e.g., a replay attack under which the real-time process data are replaced with their historic "attack-free" values [7]). To detect attacks that a passive detection scheme fails to detect, an active detection method utilizing an external intervention is used. Several approaches for active detection have been proposed. Examples include, a method using a watermarking signal to detect attacks [7], and a moving target defense under which an auxiliary system with time-varying dynamics is added to the process [8]. However, the influence of controller design on the ability (or lack thereof) of a detection method [9] for the design of an active detection method has not been explored extensively.

In this work, approaches for the design of an active detection method that leverages the influence of controller design on the ability of a detection scheme to detect an attack are presented. The detection of multiplicative and replay FDI attacks is considered. Implementing an external intervention on the attack-free process may induce transients in the process and trigger false alarms in a detection scheme. A reachable set-based detection scheme that guarantees a zero false alarm rate from an external intervention is utilized. Attacks are classified as detectable, undetectable, and potentially detectable based on the ability of the detection scheme to detect them (called attack detectability). The interdependence between controller design, the closed-loop stability of the attacked process, and attack detectability is rigorously analyzed. Leveraging the analysis, approaches for the design of an active detection method that enables attack detection by enhancing attack detectability is presented. The active detection method may use one of two different types of controller design-based interventions. Under the first intervention, a control design parameter switch is implemented to operate the process under the so-called "attack-sensitive" parameters. The attack-sensitive parameters are chosen such that an attack on the process destabilizes it, thereby enabling attack detection. However, attack-sensitive parameters may not exist for all classes of FDI attacks. To guarantee the detection of FDI attacks for which attack-sensitive parameters may not exist (e.g., replay FDI attacks), the second intervention with a setpoint change to induce the so-called "attack-revealing" perturbations in the process may be used. Using either intervention may result in undesirable performance degradation in the process, and a tradeoff between attack detection and performance degradation may exist. Approaches for managing the tradeoff through active detection method design are discussed. Finally, the design and implementation of an appropriate active detection method is demonstrated using simulations of an illustrative process example. The results demonstrate that the detection of multiplicative and replay FDI attacks may be facilitated by an active detection method utilizing an appropriate controller design-based intervention.

References:

[1] IBM., "What is Industry 4.0?", https://www.ibm.com/topics/industry-4-0. Accessed: 9th June 2024.

[2] Liang, G., Zhao, J., Luo, F., Weller, S.R., and Dong, Z.Y., "A review of false data injection attacks against modern power systems", IEEE Transactions on Smart Grid, vol. 8(4), pp.1630-1638, 2016.

[3] Mtukushe, N., Onaolapo, A.K., Aluko, A., and Dorrell, D.G., "Review of cyberattack implementation, detection, and mitigation methods in cyber-physical systems", Energies, vol. 16(13), p.5206, 2023.

[4] Aljundi, I., Rawashdeh, M., Al-Fayoumi, M., Al-Badarneh, A., and Al-Haija, Q.A., "Protecting Critical National Infrastructures: An Overview of Cyberattacks and Countermeasures", In proceedings of the international conference on WorldS4, Singapore, pp. 295-317, 21-24 August, 2023.

[5] Zhang, D., Wang, Q.G., Feng, G., Shi, Y., and Vasilakos, A.V., "A survey on attack detection, estimation and control of industrial cyber–physical systems", ISA transactions, vol. 116, pp. 1-16, 2021.

[6] Murguia, C. and Ruths, J., "CUSUM and chi-squared attack detection of compromised sensors", In proceedings of the 2016 IEEE Conference on Control Applications, pp. 474-480, Buenos Aires, Argentina, Sep. 19 - 22, 2016

[7] Liu, H., Mo, Y., and Johansson, K. H., "Active detection against replay attack: A survey on watermark design for cyber-physical systems," in Lecture Notes in Control and Information Sciences. Springer, 2021, pp. 145–171.

[8] Babadi, N. and Doustmohammadi, N., "A moving target defence approach for detecting deception attacks on cyber-physical systems," Computers and Electrical Engineering, vol. 100, p. 107931, 2022.

[9] Narasimhan, S., El-Farra, N. H., and Ellis, M. J., "Detectability-based controller design screening for processes under multiplicative cyberattacks," AIChE Journal, 68:e17430, 2022.

[10] Reda, H.T., Anwar, A. and Mahmood, A., "Comprehensive survey and taxonomies of false data injection attacks in smart grids: attack models, targets, and impacts", Renewable and Sustainable Energy Reviews, vol. 163, p.112423, 2022.

702c Quantum Federated Learning-Based Collaborative Manufacturing

David E. Bernal Neira

With the rise of the Industrial Internet of Things, the vulnerability of smart manufacturing systems to potential threats has escalated, posing significant risks to production processes, industrial assets, and sensitive data. Consequently, there is a need to actively monitor the data within industrial control systems to identify defects promptly. In this paper, we combined the principles of quantum computing with the collaborative nature of federated learning. We proposed a quantum federated learning framework, which is a new quantum machine learning framework to use the scattered manufacturing data while protecting data privacy. The proposed framework is helps in minimize maintenance costs, ensure the security of manufacturing data, and also achieve the best performance compared to the local enterprise models, regardless of how the manufacturing data is distributed among them. It combines the collaborative advantages of federated learning with the unique capabilities of quantum computing, offering a solution that is secure, scalable, and adaptable to the complexities of modern manufacturing environments.

Till now, several machine learning algorithms have been applied for defect detection in manufacturing, contributing to improved product quality and operational efficiency [1]. Indeed, the significance of high-quality, clean, well-labeled, and sufficiently large datasets in ensuring the success of machine learning models [2][3]. Stringent privacy measures may restrict access to certain data, limiting the amount of information available for model training. Stringent privacy measures may restrict access to certain data, limiting the amount of information available for model training. The sensitivity and value of manufacturing data derive from its indispensable role in modern manufacturing, its pivotal contribution to technological advancement, and its encompassment of crucial details about the manufacturing process and materials utilized. Therefore, any breach in data security poses a multifaceted risk that extends beyond immediate financial losses to encompass long-term damage to a company's reputation and competitive standing [4]. Balancing the need for data privacy with the requirements for effective machine learning in manufacturing involves adopting advanced privacy-preserving techniques and fostering collaboration within the industry while respecting legal and ethical boundaries.

In recent times, Federated Learning (FL) has transformed the research landscape, enabling model training across decentralized clients without the need for a direct exchange of raw data [5]. In the last few years, there has been a rapid increase in the adoption of Quantum Machine Learning (QML) as a framework to tackle computational challenges across several domains [6].

In this presentation, a privacy-preserving quantum federated learning framework consisting of quantum circuits to collaboratively address defect detection across diverse manufacturing units. The findings indicate that the suggested framework demonstrated a defect detection performance on par with centralized learning, which involves data sharing among manufacturers/clients but compromises data privacy. Moreover, it notably surpassed individual learning, wherein each manufacturer develops a model using its proprietary data. Furthermore, the study reveals that incorporating data diversity both within and across clients enhances the performance of quantum federated learning. Moreover, the quantum global model does not incur a substantially higher training cost compared to centralized learning.

References

1. T. Yu, G. H. Lin, I. H. R. Jiang, & C. Chiang, "Machine-learning based hotspot detection using topological classification and critical feature extraction." In Proceedings of the 50th annual Design Automation Conference, pp. 1-6, 2013.
2. Ding, J. Torres, & D. Pan, "High-Performance lithography hotspot detection with successively refined pattern identifications and machine learning." IEEE Transactions On Computer-Aided Design Of Integrated Circuits And Systems. 30, 1621-1634 (2011).
3. Shin & J. Lee, "Accurate lithography hotspot detection using deep convolutional neural networks". Journal of Micro/Nanolithography, MEMS, And MOEMS. 15, 043507-043507 (2016)

4. Mehta, M., & Shao, C. (2022). Federated learning-based semantic segmentation for pixel-wise defect detection in additive manufacturing. Journal of Manufacturing Systems, 64, 197-210.
5. McMahan, B., Moore, E., Ramage, D., Hampson, S. & Arcas, B. Communication-efficient learning of deep networks from decentralized data. Artificial Intelligence And Statistics. pp. 1273-1282 (2017)
6. Biamonte, J., Wittek, P., Pancotti, N., Rebentrost, P., Wiebe, N., & Lloyd, S. (2017). Quantum machine learning. Nature, 549(7671), 195-202.

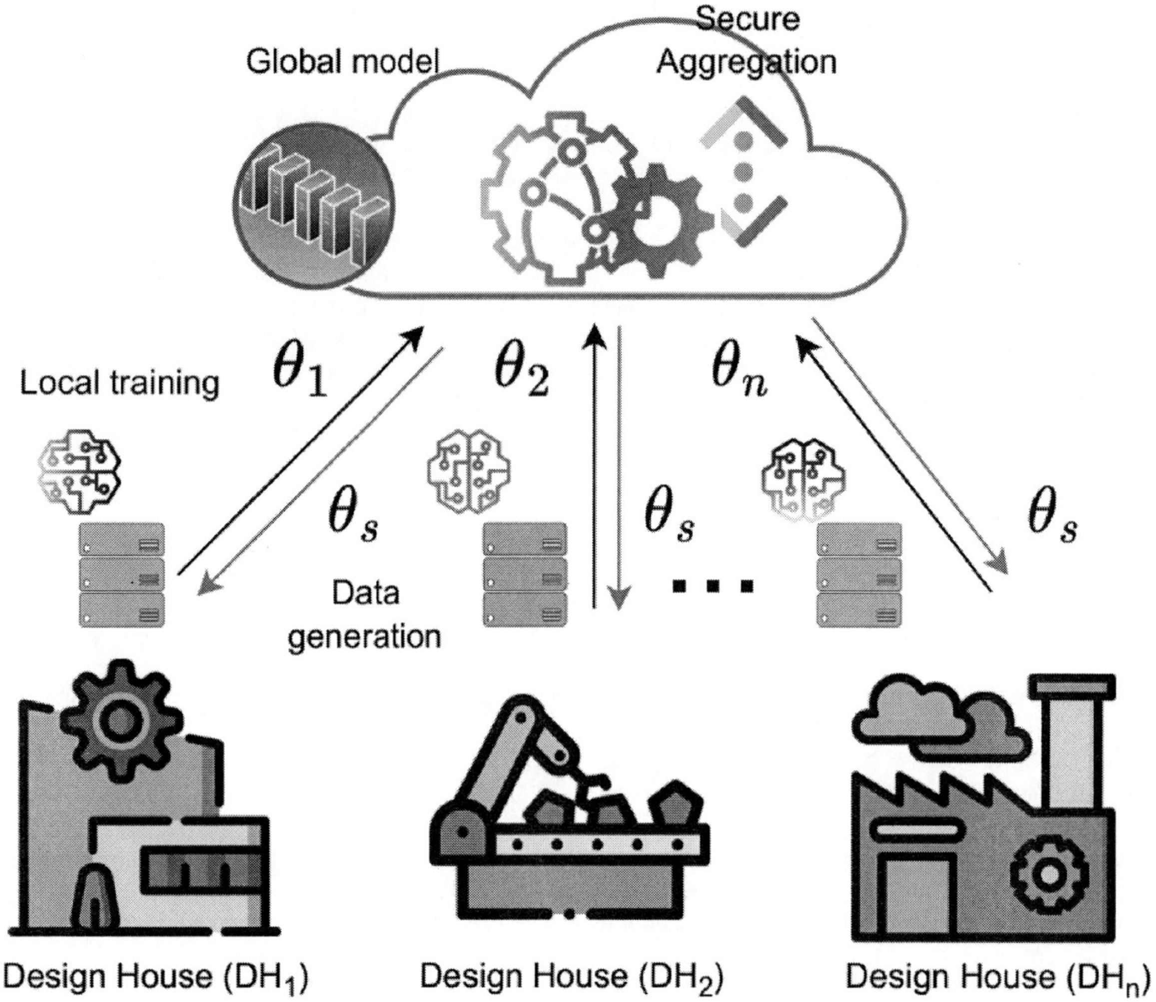

702d Cyberattack Detection and Handling for Neural Network-Approximated Economic Model Predictive Control

Jihan Abou Halloun, Helen Durand

Cyberattacks on control systems can create unprofitable and unsafe operating conditions. A common approach to enhance safety and attack resiliency of control systems is to develop attack detection mechanisms. In prior work in our group, a number of cyberattack detection strategies were elaborated for chemical processes integrated with an advanced control formulation known as Lyapunov-based economic model predictive control (LEMPC) [1], in the sense that the controller properties can be used to analyze closed-loop stability in the presence or absence of undetected attacks [2,3]. This raises two questions of whether these strategies are more conservative than alternative detection methodologies such as those based on reachability analysis for attack detection (e.g., [4], [5]) and also to what extent integrating detection strategies with LEMPC alone is useful.

In this work, we move toward addressing these questions [7]. First, we demonstrate that the LEMPC-based cyberattack detection strategies can be modified to enable extension to a neural network (NN) version of the controller (i.e., using a NN as a function approximator [6]). This NN then approximates the LEMPC (which will be abbreviated as NN-LEMPC). We provide sufficient conditions for an LEMPC which ensure that if an NN-LEMPC is then designed based on data from that LEMPC, the NN-LEMPC will inherit safety properties from the original LEMPC formulation. We focus on this for one of the passive detection strategies for LEMPC from [2], in which a threshold on the difference between state predictions and estimates is used to detect attacks. Then, we discuss a second detection and handling strategy based on reachability analysis and input repair (inspired by the concept of repairing a NN in [8]) for ensuring safety of controllers (e.g., EMPC) that are not based on a rigorous control law formulation like LEMPC (in the sense of boundedness of the closed-loop state in a safe operating region) for a sampling period after an undetected cyberattack. We use the same detection metric related to state predictions and estimates to facilitate a comparison between the LEMPC-based methods and the reachable set and input repair technique. This enables use to examine the potential conservatism differences between the LEMPC-based safety strategy and the detection and handling concept based on repairing problematic control actions that correspond to false state measurements due to a sensor attack, and faciliates discussion regarding how this concept can inspire ideas for fighting back against attacks.

References:

[1] Heidarinejad, Mohsen, Jinfeng Liu, and Panagiotis D. Christofides. 'Economic model predictive control of nonlinear process systems using Lyapunov techniques.' *AIChE Journal* 58, no. 3 (2012): 855-870.

[2] Oyama, Henrique, and Helen Durand. 'Integrated cyberattack detection and resilient control strategies using Lyapunov-based economic model predictive control.' *AIChE Journal* 66, no. 12 (2020): e17084.

[3] Oyama, Henrique, Dominic Messina, Keshav Kasturi Rangan, and Helen Durand. 'Lyapunov-based economic model predictive control for detecting and handling actuator and simultaneous sensor/actuator cyberattacks on process control systems.' *Frontiers in Chemical Engineering* 4 (2022): 810129.

[4] Liu, Hao, Ben Niu, and Jiahu Qin. 'Reachability analysis for linear discrete-time systems under stealthy cyber attacks.' *IEEE Transactions on Automatic Control* 66, no. 9 (2021): 4444-4451.

[5] Narasimhan, S., El-Farra, N.H. and Ellis, M.J., 2023. A reachable set-based scheme for the detection of false data injection cyberattacks on dynamic processes. *Digital Chemical Engineering*, *7*, p.100100.

[6] Lucia, Sergio, and Benjamin Karg. 'A deep learning-based approach to robust nonlinear model predictive control.' *IFAC-PapersOnLine* 51, no. 20 (2018): 511-516.

[7] Abou Halloun, Jihan and Helen Durand. 'Cyberattack Detection and Handling for Neural Network-Approximated Economic Model Predictive Control.' *ADCHEM 2024* (2024), accepted.

[8] Yang, Xiaodong, Tom Yamaguchi, Hoang-Dung Tran, Bardh Hoxha, Taylor T. Johnson, and Danil Prokhorov. 'Neural network repair with reachability analysis.' In *International Conference on Formal Modeling and Analysis of Timed Systems*, pp. 221-236. Cham: Springer International Publishing, 2022.

702e A Methodology for Control Configuration Screening for Cyberattack Identification in Process Systems

Aatam Gajjar, Matthew Ellis, Nael El-Farra

Process Control Systems (PCS) represent a class of cyber-physical systems that integrate physical processes, computational resources, and communication capabilities. Recent incidents of various industrial security breaches have highlighted the vulnerability of cyber-physical systems to cyberattacks, particularly targeting their data management and communication layers [1]. Among the various attack vectors, one particularly concerning avenue is the infiltration of communication links within PCS. Specifically, cyber-attackers may gain access to the control system network and then alter the data communicated over the sensor-controller or the controller-actuator links, resulting in performance degradation or, worse, instability [1, 2, 3]. While cyberattack threats to PCS have traditionally been treated as information technology (IT) issues and addressed through IT infrastructure-based strategies, the increased frequency of cyberattacks has revealed the inadequacy of such approaches and the need for complementary approaches utilizing operational technology (OT)-based strategies. Consequently, OT-based approaches for the design of PCS that are resilient to cyberattacks have received increased attention recently [3].

OT-based approaches to the design of cyber-resilient PCS require the development of attack detection and identification schemes. Attack detection schemes aim to detect the presence of an attack by identifying abnormal operating behavior from normal operations [4]. Different strategies have been proposed for attack detection, including various data-based approaches (see [2, 3, 4] for recent surveys). Moreover, the influence of control system parameters on the ability of a detection scheme to detect an attack has been established [5] and used as the basis for the development of active attack detection schemes [6].

Attack identification schemes, on the other hand, are aimed at revealing the compromised components of the system, isolating the attacks from other disturbances, and/or estimating the severity of the attacks [7]. Results on the identification of cyberattacks in PCS are limited at present, especially when compared with the extensive body of research work on the problem of attack detection. Examples of existing contributions include the development of data-based (e.g., [8, 9]) and model-based methods (e.g., [10]) for attack isolation.

While exploiting process model knowledge for attack identification is justified in light of the increasing use of model-based controllers, the ability of model-based attack identification schemes to distinguish between attacks and other disturbances, or between different kinds of attacks, rests fundamentally on the underlying process structure captured by the process model. This structure determines the channels through which the various process inputs influence the process outputs. In this context, the selection of the control system structure has the potential to play an important role in enabling the isolation of certain attacks by capitalizing on the underlying process structure. At this stage, however, a rigorous characterization of the connection between the choice of the control system structure and the ability of an identification scheme to isolate attacks remains unexplored. The current work aims to bridge this gap.

Motivated by these considerations, we focus in the present work on the problem of attack isolation in processes subject to false-data injection attacks that manipulate the data transmitted over the controller-actuator links. We consider a model-based attack isolation scheme that utilizes a set of unknown input observers with dedicated residuals to isolate the attacks from process disturbances and

distinguish different types of attacks. Conditions that
guarantee attack isolation are obtained and used to establish the fundamental connection between the process and control system structures, on the one hand, and the ability of the identification scheme to isolate the attacks, on the other. This characterization serves as a foundation for the development of a screening methodology that determines whether certain attacks may or may not be isolated for a given choice of the control configuration. The proposed methodology integrates attack identification as an additional criterion in the selection of the control system configuration. The results are illustrated using a chemical process example.

References

[1] T. Alladi, V. Chamola, and S. Zeadally, "Industrial control systems: Cyberattack trends and countermeasures," Computer Communications, vol. 155, pp. 1–8, 2020.

[2] N. Mtukushe, A. K. Onaolapo, A. Aluko, and D. G. Dorrell, "Review of cyberattack implementation, detection, and mitigation methods in cyber-physical systems," Energies, vol. 16, p. 5206, 2023.

[3] S. Parker, Z. Wu, and P. D. Christofides, "Cybersecurity in process control, operations, and supply chain," Computers & Chemical Engineering, p. 108169, 2023.

[4] D. Zhang, Q.-G. Wang, G. Feng, Y. Shi, and A. V. Vasilakos, "A survey on attack detection, estimation and control of industrial cyber–physical systems," ISA transactions, vol. 116, pp. 1–16, 2021.

[5] S. Narasimhan, N. H. El-Farra, and M. J. Ellis, "Detectability-based controller design screening for processes under multiplicative cyberattacks," AIChE Journal, vol. 68, no. 1, p. e17430, 2022.

[6] ——, "Active multiplicative cyberattack detection utilizing controller switching for process systems," Journal of Process Control, vol. 116, pp. 64–79, 2022.

[7] F. Pasqualetti, F. Dörfler, and F. Bullo, "Attack detection and identification in cyber-physical systems," IEEE transactions on automatic control, vol. 58, pp. 2715–2729, 2013.

[8] Z. Wu, F. Albalawi, J. Zhang, Z. Zhang, H. Durand, and P. D. Christofides, "Detecting and handling cyber-attacks in model predictive control of chemical processes," Mathematics, vol. 6, no. 10, p. 173, 2018.

[9] S. Chen, Z. Wu, and P. D. Christofides, "Cyber-attack detection and resilient operation of nonlinear processes under economic model predictive control," Computers & Chemical Engineering, vol. 136, p. 106806, 2020.

[10] M. Kordestani and M. Saif, "Observer-based attack detection and mitigation for cyber-physical systems: A review," IEEE Systems, Man, and Cybernetics Magazine, vol. 7, pp. 35–60, 2021.

702f Extending Lyapunov-Based Active Cyberattack Detection to Terminal Equality-Constrained Control and Directed Randomization to Quantum Control

Jihan Abou Halloun, Kip Nieman, Keshav Kasturi Rangan, Helen Durand

Cybersecurity considerations are becoming more critical due to the high exposure of control systems to a range of threats and security vulnerabilities. To enhance safety and attack resiliency of control systems, passive cyberattack detection strategies (e.g., [1]) as well as active detection methodologies (e.g., [2]) can be developed. In our prior group's work (i.e., [3]), an active detection strategy for chemical processes under Lyapunov-based economic model predictive control (LEMPC) [4] was elaborated. This strategy specific to LEMPC probed for cyberattacks by using random modifications of the LEMPC generated at random times to check whether the modified Lyapunov functions decreased over the subsequent sampling period (which should be the case in the absence of an attack). However, though LEMPC was used for cyberattack detection policies in our prior works, LEMPC is not the only EMPC framework. Thus, it is important to understand to what extent the insights revealed by studying the cyberattack detection policies within an LEMPC-based framework translate to other EMPC formulations (e.g., a terminal equality constraint formulation [5]). In addition to LEMPC-focused cyberattack detection strategies, our prior work has considered another strategy intended for the case that all sensors are under attack. We termed this strategy "directed randomization" because randomly selects between two control actions at every sampling time of process operation [6]. This is done to attempt to make it hard for an attacker that is targeting the sensors to consistently predict which control action should have been applied to match their false state measurements to it. Randomness plays an important role in this strategy; this therefore raises another question regarding whether such detection strategies can be extended to the control of systems with inherent randomness, in particular quantum systems.

Motivated by these considerations, this talk will be divided into two parts. In the first part, we first show how the key components of the active detection policy developed in [3] can be extended to a terminal equality constraint EMPC formulation. We discuss the role of the prediction horizon length in our ability to extend the detection concepts from [3], and provide the conditions under which safety and feasibility of this strategy are maintained in the absence as well as in the presence of an attack on the sensors for one sampling period. In the second part of this talk, we investigate the challenges with applying "directed randomization" to the mitigation of cyberattacks on quantum systems. We first provide a background on quantum mechanics and quantum control where we focus specifically on the use of a laser to control the bond length of a hydrogen fluoride molecule [7]. We then discuss how the measurement principles of quantum mechanics can cause projective measurement of a quantum system to be challenging to integrate with directed randomization, but instead discuss how principles from [8] (in particular, attempts to design quantum algorithms which significantly reduce the probability of measuring certain outcomes) might be used to attempt to re-gain properties like those of directed randomization in probing for cyberattacks on quantum systems.

References:

[1] Oyama, Henrique, Dominic Messina, Keshav Kasturi Rangan, and Helen Durand. 'Lyapunov-based economic model predictive control for detecting and handling actuator and simultaneous sensor/actuator cyberattacks on process control systems.' *Frontiers in Chemical Engineering* 4 (2022): 810129.

[2] Narasimhan, Shilpa, Nael H. El-Farra, and Matthew J. Ellis. 'Active multiplicative cyberattack

detection utilizing controller switching for process systems.' *Journal of Process Control* 116 (2022): 64-79.

[3] Oyama, Henrique, and Helen Durand. 'Integrated cyberattack detection and resilient control strategies using Lyapunov-based economic model predictive control.' *AIChE Journal* 66, no. 12 (2020): e17084.

[4] Heidarinejad, Mohsen, Jinfeng Liu, and Panagiotis D. Christofides. 'Economic model predictive control of nonlinear process systems using Lyapunov techniques.' *AIChE Journal* 58, no. 3 (2012): 855-870.

[5] Amrit, Rishi, James B. Rawlings, and David Angeli. 'Economic optimization using model predictive control with a terminal cost.' *Annual Reviews in Control* 35, no. 2 (2011): 178-186.

[6] Oyama, Henrique, Dominic Messina, Keshav Kasturi Rangan, Akkarakaran Francis Leonard, Kip Nieman, Helen Durand, Katie Tyrrell, Katrina Hinzman, and Michael Williamson. 'Development of directed randomization for discussing a minimal security architecture.' Digital Chemical Engineering 6 (2023): 100065.

[7] Magann, Alicia B., Matthew D. Grace, Herschel A. Rabitz, and Mohan Sarovar. 'Digital quantum simulation of molecular dynamics and control.' *Physical Review Research* 3, no. 2 (2021): 023165.

[8] Nieman, Kip and Helen Durand. "Safety with Non-Deterministic Control Action Selection Using Quantum Devices." *ADCHEM 2024* (2024), accepted.

234a Printability Criterion of Highly Filled Pastes for Direct-Ink Writing Based on Small-Amplitude Oscillatory Shear (SAOS) Experiments and Packing Fraction

Jessica Kopatz, James Griebler, Jonathan Leonard, Alexander Tappan, Anne Grillet

Direct-ink write is an additive manufacturing technique that enables the creation of reproducible and complex hardware by depositing a viscous, shear-thinning liquid onto a substrate in a custom-pattern via extrusion through a syringe. The rheology of these inks is tailored through the addition of various filler materials. To successfully print highly-filled inks, we need to understand the effect of filler morphology, size, loading, and packing fraction on the ink rheology and corresponding printability. More importantly, characterization methods that accurately capture the ink's rheological properties that correlate to printability is imperative. We present results which demonstrate why traditional characterization of these systems through steady-shear tests are difficult and often unreliable. We then offer a more reliable methodology to determine printability. Various filler particles and volume loadings of particles were dispersed in Polydimethylsiloxane (PDMS, Sylgard® 182) and extruded through a syringe to determine the printability of each material separated into three categories: Thin, Printable, and Thick. We then correlate measurements made through small-amplitude oscillatory shear to the printability of each finished product and show that some of the common metrics (linear regime storage modulus and complex viscosity) as well as often ignored metrics (axial force) can distinguish between materials which are printable and materials which are too thick to be extruded. We calculate the rate of change of the strain amplitude and axial force in time, to see how fast materials are changing behavior, which we have paired with images which show that the axial force drops due to material leaving the gap between plates. Finally, we used the linear regime loss modulus, which is associated with energy dissipation, to fit the Krieger-Dougherty to determine the maximum packing fraction of each particle. In doing so we have determined a printability criterion based on packing fraction normalized by the maximum packing fraction to separate materials which are printable from those that are too thick or too thin.

Sandia National Laboratories is a multi-mission laboratory managed and operated by National Technology and Engineering Solutions of Sandia, LLC., a wholly owned subsidiary of Honeywell International, Inc., for the U.S. Department of Energy's National Nuclear Security Administration under contract DE-NA0003525.

234g Large Language Models Monitor and Control Additive Manufacturing

Amir Barati Farimani

Industry 4.0 has transformed manufacturing, particularly through additive manufacturing (AM) like Fused Deposition Modeling (FDM), which enables cost-effective, customized production with minimal waste. However, FDM's error susceptibility often requires expert oversight to detect and correct defects that compromise quality. Existing automated solutions lack adaptability across different printers, and deep learning models demand large labeled datasets, limiting scalability. To overcome this, we propose a process monitoring framework using pre-trained Large Language Models (LLMs) to detect and address defects in real-time. Validations show that LLM-based agents can autonomously correct common 3D printing issues, matching the performance of experienced engineers.

234c Generalizing Cell Culture Models from Fed-Batch to Continuous Using Metabolic Knowledge

Mariana Monteiro, Stacy Bediako, Gregory Nierode, Pavlos Kotidis, Cleo Kontoravdi

Continuous biomanufacturing has the potential to significantly enhance the yield, productivity, and consistency of therapeutic protein production. However, the transition of cell lines which are typically optimized for fed-batch operation to continuous mode presents a challenge, particularly in selecting *a priori* satisfactory continuous operating parameters.

Current methodologies rely on extensive experimentation, which is time- and cost-intensive, even when aided by design of experiments. This motivates the need for model-based methods that can reduce the experimental load by predicting performance, based on existing data and prior knowledge, leveraging experimentation only for final validation. Existing models, however, are overparameterized to experimental conditions, thus not generalizable across cell lines, feeding strategies and other operating parameters. It is possible to increase a model's generality, but this requires more data and thus more experimentation, which ultimately incurs the same costs. Neither of these approaches are satisfactory as they do not address the underlying problem of requiring extensive and costly experimentation.

We propose overcoming the need for extensive experimentation by building more general reactor models through embedding first-principles knowledge. By incorporating cell metabolism in a nonparametric model of the production reactor, we are able to predict key process parameters like metabolite concentrations without requiring kinetic parametrisation. By limiting ourselves to the core metabolic reactions common across different mammalian cells, our novel multi-scale formulation offers increased predictive capability and generality. Additionally, by introducing mode and scale-specific constraints, our formulation can predict across different modes of operation without requiring kinetic parameter estimations, thereby reducing the need for extensive experimental data.

We present an evaluation of our model by comparing its predictions against both fed-batch and continuous industrial cell culture data. This novel formulation can be used to estimate performance for transitioning from fed-batch to continuous, by incorporating mode-specific constraints. The applicability of this formulation is extended to control, namely soft sensing of hard-to-measure metabolites and to estimate when to finish the culture optimally, further demonstrating the versatility of the proposed model.

234d Virtual Testing of Image Sensors for Equipment and Operator Safety

Akkarakaran Leonard, Muhammad Adnan, Helen Durand

The rise of automation has lead to an increase in tasks being handled by machines as opposed to human operators. However, operators perform an essential role in the running of modern plants, performing tasks such as fault detection, supervision of overall functioning of the plant, and being the human-in-the-loop for large automated processes. This has lead to fewer operators that handle some of the most sensitive tasks in a plant's day to day operation. Image sensors can be a versatile method of maintaining safety, for example, in automotive applications using CMOS image sensors, as in the works of Elkhalili et al. [1] where the sensors are used to determine the distance of objects in the visual field, and in Yamazato et al. [2], which uses the sensors to carry out visual light communication (VLC) to facilitate communication between individual vehicles. The safety of operators in Chemical Engineering has been explored through the modeling of operator behavior [3], and the analysis of operator training [4]. In this paper we demonstrate the use of image sensors as an automated means of maintaining the safety of human operators. These image sensors are simulated virtually using Blender, a free and open source 3D modeling and animation software. A model of a human operator connecting and disconnecting a hose to a tank in a room are created in Blender, and a camera is set up to capture images of the operator, the hose connector, and operators hands as they perform the task. Here we develop three cases of use for image sensors in safety applications: (1) to determine the position of the operator based on images captured by the camera, to function as a location-based safety alert system. (2) to determine the position of the hose connector from the outlet of the tank, to detect if the connector is an unsafe position. (3) to determine if the human operator performs the task in a safe manner by analyzing the haptic information of the operator observed by the camera.

References

[1] Omar Elkhalili et al. "A 64× 8 pixel 3-D CMOS time of flight image sensor for car safety applications". In: 2006 Proceedings of the 32nd European Solid-State Circuits Conference. IEEE. 2006, pp. 568–571.

[2] Takaya Yamazato et al. "Image-sensor-based visible light communication for automotive applications". In: IEEE Communications Magazine 52.7 (2014), pp. 88–97.

[3] YM Sebzali and XZ Wang. "Joint analysis of process and operator performance in chemical process operational safety". In: Journal of Loss Prevention in the Process Industries 15.6 (2002), pp. 555–56

[4] John Haesle, Chris Devlin, and Jack L Mccavit. "Improving process safety by addressing the human element". In: Process Safety Progress 28.4 (2009), pp. 325–330.

234f Statistical Analysis of Processes Using Data Provided By the PI System (AVEVA/OSISoft)

Waleska Nascimento, Gabriela Monteiro, Heleno Bispo, Fernando Schramm

The process industry digital transformation, through the adoption of Industry 4.0 technologies, plays a significant role in optimization, cost reduction, and the improvement of the quality/performance of the final product and processes (Fogler, 1999; Roffel e Betlem, 2006). Therefore, the application of statistical methods, such as Statistical Process Control (SPC), contributes to ensuring the quality and efficiency of processes, leading to operational excellence. SPC aims to quickly detect variations in the process, preventing further losses to the company (Montgomery, 2009). Furthermore, to enhance process efficiency, improve decision-making, generate insights for new solutions, and increase business profitability, an increasing number of companies are resorting to the use of predictive models, taking advantage of the union of these topics.

In this context, the use of software that has the capacity to store and provide access to a large amount of data, collected in real time, for use in the development of the statistical analysis environment in Python and the predictive model is mandatory. Then, the PI System, software from AVEVA/OSIsoft, emerges as a valuable tool for this process, providing capabilities for data acquisition, visualization, and analysis. On this way, the main objective of this project is the development of a Python-based tool for statistical analysis of processes, using data from the PI System (AVEVA/OSIsoft), and expanding its usability through a predictive model for quality control based on Statistical Process Control (SPC).

The methodology used integrates a module written and executed in a Python environment and with an OPC communication, ensuring connectivity with the PI System database and visualization applications (Figure 1). This structure allowed the development of supervisory interfaces for monitoring processes and decision-making. For the predictive model, the CRISP-DM methodology, that is widely used in Data Science projects (Piatetsky, 2014), was applied. Such a methodology encompasses six stages: business understanding, data understanding, data preparation, modeling, evaluation, and deployment. In this project, the focus was on the data preparation stage, which involved extracting data from the PI System, exporting it to a CSV file, and ensuring its compatibility with analysis tools. This prepared data was then analyzed in Power BI, utilizing its advanced visualization and analysis capabilities (Figure 2).

Figure 1 - PI System architecture for data collection, storage and delivery
Figure 2 - Exploratory data analysis in Power BI

The results highlight the potential of integrating SPC and the PI System in the simulation of industrial events and in the training of engineers and operators, enhancing process efficiency. Additionally, the digital transformation structure, which includes dynamic simulation software connected to the PI System, serves as a foundation for studies, covering the entire control and optimization structure, and providing a simulated environment close to a real process.

References

Fogler, H., Elements of Chemical Reaction Engineering, Prentice Hall, 1999.

Montgomery, D., Introduction to Statistical Quality Control, JohnWiley and Sons, 2009.

Piatetsky, G. Crisp-dm, still the top methodology for analytics, data mining, or data science projects. KDD News, 2014.

Roffel, B. E Betlem, B., Process Dynamics and Control - Modeling for Control and Prediction, John Wiley & Sons, 2006.

Figure 1 - PI System architecture for data collection, storage and delivery

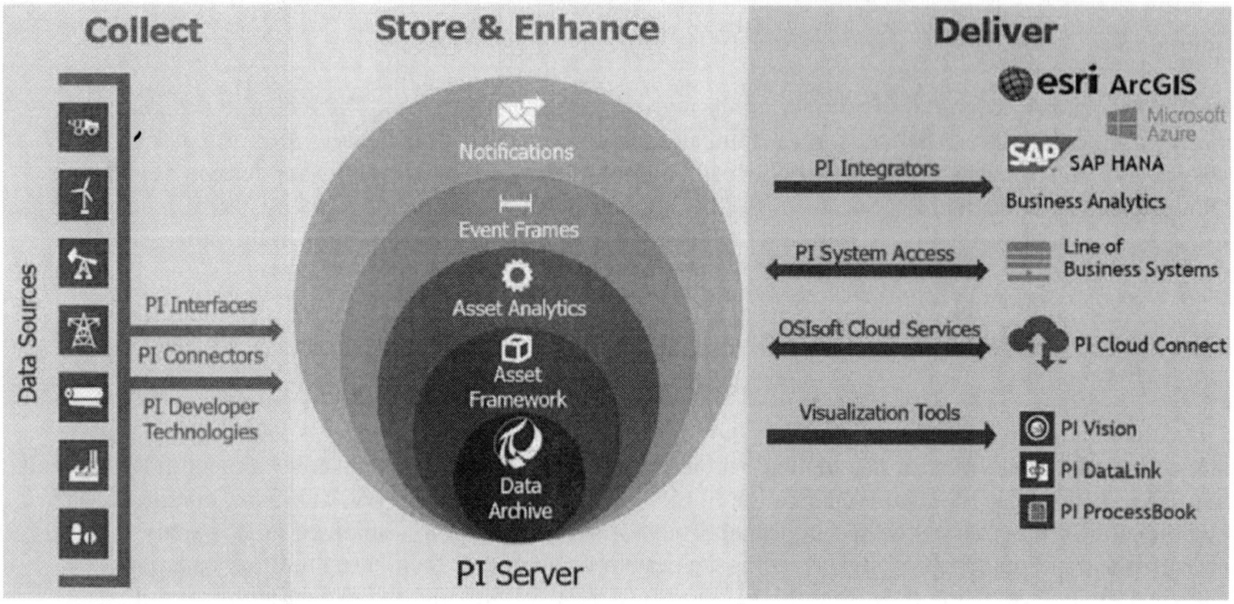

Figure 2 - Exploratory data analysis in Power BI

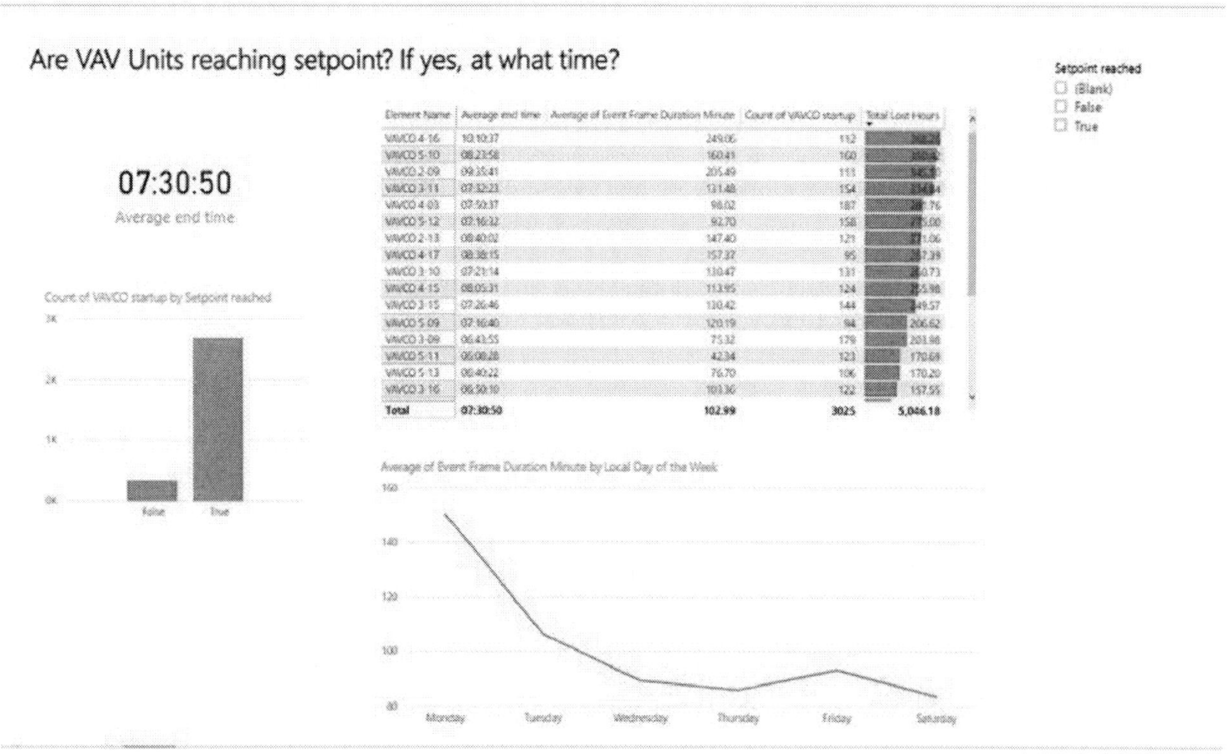

164a Accelerated Coatings Development with High Throughput Technology and Digital Infrastructure

Shashank Kamdar, Lidaris San Miguel Rivera, Sun Hye Kim, Partha Majumdar, Ben Caderaio, Jonathan Moore, Sukrit Mukhopadhyay, Michael Tran, Kevin Henderson, Jonathan DeRocher, Sunny Wang, Xiaoshuang (June) Wei, Wenzhao Yang, Sharon Vuong, Sam Lim, Alicia Liew, Nipun Bisht, Dan Wu

While the concept and use of high throughput experimentation and automation is prevalent in pharmaceutical industries, the materials industry is still in its nascent stage and faces unique challenges. These challenges are especially onerous when dealing with complex formulations such as paints and coatings that comprise of many inorganic and macromolecular materials. Furthermore, collecting relevant experimental data to describe formulations is an exhaustive, time-consuming process, and even after the large data collection, a deficient data infrastructure will hinder data analysis and subsequent insights. This talk will provide an overview of Dow Coating Materials' transformational journey towards creating, handling, and storing data more efficiently. This transformation has allowed seamless integration of data with various machine learning and advanced statistics techniques to deploy new digital applications for data driven product development.

164b Digital Twins for Continuous Manufacturing in Pharma: The Case of Dissolved Oxygen Control

Christoforos Somarakis, Roberto Irizarry, Mihai G. Buciu

The Continuous Manufacturing (CM) process provides a self-contained, single-use, reconfigurable platform, that incorporates principles of automated and continuous bioprocessing. For maintaining a steady state of biomass, effective process monitoring, and controls must be in place to ensure accurate cell density, critical nutrients and metabolites, as well as the relationship between growth and metabolic data. Measurable quantities that are typically leveraged are pH, temperature, and dissolved oxygen (DO) concentration.

An original digital twin platform is proposed for plug-and-play interaction with the existing closed loop regulation of DO in the bioreactor. It performs critical parameters' identification and enables synthesis of advanced control strategies for improved regulation of DO concentration levels. The platform is designed to cope with real-world challenges at the manufacturing level, that include but are not limited to corrupted data, unmeasurable disturbances, insufficient knowledge of process dynamics, modelling gaps and limited space for experimental designs. The platform consists of a blend of computation and processing components, that perform sequential calculations for on-the-fly, rapid learning, and system identification. It performs joint system identification and advanced control with respect to regulation of process of interest.

Our case study considers a fictitious problem of regulation of DO concentration levels in bioprocess setup that mimics Merck's CM production line. We demonstrate how the proposed digital twin architecture combines heterogeneous data from the established closed-loop process, informs the mass balance model that dynamic DO concentration follows. The identification part provides estimates of constant (e.g. kLa coefficients, time-delays etc.) as well as non-constant parameters, such as specific oxygen uptake rate. System identification is essentially an intermediate step of an inverse engineering task that delivers an informed digital twin model. The latter model will then be the basis to design model reference advanced controls. We show that the proposed Digitally Twin-enabled joint dynamic controller outperforms the existing PID control structures.

164c Enhancing Human Performance with Human-Centric Hmis in mRNA Manufacturing

Mohammed Aatif Shahab, Richard Braatz

The integration of digital twin technology in the biopharmaceutical sector, especially mRNA manufacturing, is revolutionizing manufacturing processes by offering insights into real-time operations and enabling predictive modeling (Scheper et al., 2021). However, this advancement also introduces significant complexities for human operators, who play a critical role in monitoring and intervening in the manufacturing process. These operators are further tasked with upholding the stringent quality and safety standards mandated by Good Manufacturing Practices (GMP). Human errors account for approximately 50% of all quality incidents (Wilson et al., 2015) and related problems within the pharmaceutical industry. Often, these errors are linked to complex automation strategies that overload operators with information and use sophisticated Human-Machine Interfaces (HMIs) beyond their capacity, complicating their understanding of processes (Nazir et al., 2014). Therefore, it became imperative to enhance human performance in biopharma manufacturing (Wismer et al., 2021). This work aims to develop a human-centric HMI for the continuous mRNA process, based on human factors principles, to significantly improve operator performance. By doing so, we intend to address the dual challenges of system complexity and human error.

The development of the human-centric HMI employs a multifaceted strategy, aligned with human factors principles to optimize the operator's interaction with the technology. First, we identify the operator's tasks and cognitive workflows in the continuous mRNA process to identify areas where digital twin technology can most effectively reduce cognitive load and enhance decision-making. This involves mapping out the entire mRNA process (including upstream and downstream processes) to pinpoint where real-time data and predictive analytics can provide significant benefits. Subsequently, we design an intuitive Human-Machine Interface (HMI) that serves as the primary interaction point between operators and the digital twin. This HMI is designed to present data in an accessible and understandable format, utilizing visual aids, simplified dashboards, and actionable alerts that guide operators through the decision-making process. The interface design adheres to established ergonomic and cognitive engineering principles. By integrating these strategies, the proposed human-centric HMI aims to minimize the risk of human error while maximizing operational efficiency and adherence to GMP standards. Moreover, it can serve as a testbed for future human factors research in biopharma and a tool for operator training (Wismer et al., 2021).

Keywords: mRNA, human error, biopharma, human-machine interface, digital twin

Acknowledgments: This research was supported by the U.S. Food and Drug Administration under the FDA BAA-22-00123 program, Award Number 75F40122C00200.

References

Scheper, T., Beutel, S., McGuinness, N., Heiden, S., Oldiges, M., Lammers, F., & Reardon, K. F. (2021). Digitalization and bioprocessing: Promises and challenges. *Digital Twins: Tools and Concepts for Smart Biomanufacturing*, 57-69.

Herwig, C., Pörtner, R., & Möller, J. (Eds.). (2021). *Digital Twins: tools and concepts for smart biomanufacturing*. Cham: Springer International Publishing.

Wilson, A., Moedler, M., & McAuley, G. (2015). Changing the Performance Paradigm in Pharma/Biotech: Integrating Human Performance in Global Organizations. *PDA Journal of Pharmaceutical Science and Technology*, *69*(5), 658-665.

Wismer, P., Lopez Cordoba, A., Baceviciute, S., Clauson-Kaas, F., & Sommer, M. O. A. (2021). Immersive virtual reality as a competitive training strategy for the biopharma industry. *Nature Biotechnology*, *39*(1), 116-119.

164d Industry 4.0: Artificial Intelligence (AI) and Digital Transformation of Advanced Continuous Pharmaceutical Manufacturing

Ravendra Singh

Currently, Industry 4.0 concepts are being applied to pharma industry to achieve Pharma 4.0 paradigm. Pharma 4.0 reduces the time and resources needed for continuous pharmaceutical manufacturing and also improves the product quality and production consistency. It has many advantages but also have bigger challenges on the applications of artificial intelligence (AI)/machine learning (ML), material traceability, optimization, advanced process control, cyber-physical security, and data management side because of the different levels of complexities involved [1-4]. The predictive capabilities and the quality of the pharmaceutical products can be improved significantly via employing the artificial intelligence and the advanced model predictive control (MPC) system if an appropriate cyber-physical security defense is in place.

In this work, seven components of industry 4.0 namely artificial intelligence (AI)/machine learning (ML), modelling, material traceability, optimization, advanced control, cyber-physical security, and data science have been developed and implemented into the continuous pharmaceutical manufacturing process.

Four machine learning (ML) models have been trained to predict the response of continuous pharmaceutical manufacturing process and the performance of these ML models has been compared. The investigated ML methods are long short-term memory (LSTM), 1D convolution neural network (CNN), random forest (RF), and artificial neural network (ANN). The best performing ML model is then implemented into the continuous pharmaceutical tablet manufacturing process for real time prediction. A digital twin model of continuous pharmaceutical tablet manufacturing process has been also developed.

A systematic framework including the methods and tools have been developed for material traceability. A corresponding software tool has been also developed to automate the material traceability procedure. The heart of the material traceability toolbox is the RTD model. The applications of the developed methods and tools have been demonstrated for material traceability of continuous pharmaceutical manufacturing process.

A systematic framework including the methods and tools has been developed for dynamic optimization of the feeder refill strategies. The deviation of the outlet mass flow of the feeder from the targeted flow rate has been minimized to obtain the optimum value of the feeder refill parameters. The material properties also affect the refill strategy meaning that the feeder refill strategy need to be frequently optimized if there are any changes in the materials and plant. Therefore, the developed feeder model and dynamic optimization tool can save the time and recourses as well as can improve the product quality significantly.

An advanced model predictive control (MPC) system has been implemented in the continuous pharmaceutical manufacturing (CPM) pilot-plant. The CPP's and CQA's are controlled in real time. The critical control variables that have been controlled using model predictive control (MPC) system are drug concertation, powder level before tablet press, main and pre compression forces, tablet weight and hardness. A novel control strategy for powder level control in a chute placed in between blender and tablet press unit operation of continuous tablet manufacturing process has been developed, implemented, and evaluated. An RTD tool box has been developed to control the tablet potency in real time.

A systematic framework including the methods and tools have been also developed for proactive identification and mitigation of potential cyber-physical attack risk on continuous pharmaceutical manufacturing plant [2]. The cyber-physical security relevant software tools such as Snap 7, Wireshark, and Tripwire have been applied to CPM. A novel software tool named CPS (Cyber-Physical Security) has been developed for cyber-physical security of the continuous pharmaceutical manufacturing. The integrated commercially available and developed (in house) cyber-physical security tools have added an extra layer of security of our continuous pharmaceutical manufacturing pilot-plant for any unexpected attacks.

All the relevant data generated during continuous manufacturing has been systematically collected, stored and organized in a data hub (OSI PI) and cloud system as per industry 4.0 standard.

References

1. Bhaskar, A., Barros, F. N., Singh, R. (2017). Development and implementation of an advanced model predictive control system into continuous pharmaceutical tablet compaction process. International Journal of Pharmaceutics, 534 (1-2), 159-178.
2. Singh, R. (2023). Pharma 4.0: Advanced continuous pharmaceutical tablet manufacturing. Pharma Focus Issue 2. https://issuu.com/verticaltalk/docs/pfe-issue-02/s/27554699
3. Billups, M., Singh, R. (2018). Systematic framework for implementation of material traceability into continuous pharmaceutical tablet manufacturing process. Journal of Pharmaceutical Innovation. https://doi.org/10.1007/s12247-018-9362-9.
4. Singh, R. (2021). Cyber-Physical Security (CPS) Tool for Continuous Pharmaceutical Manufacturing Process. Pharma Focus Asia. Issue 44, 41-49. https://www.pharmafocusasia.com/information-technology/cyber-physical-security-tool.

164e Enhancing Extrapolation Capabilities of a Data-Driven Prediction Model for an Air Separation Unit Using a Digital Twin

Valentin Krespach, Nicolas Blum, Martin Pottmann, Sebastian Rehfeldt, Harald Klein

In air separation units, improvement of plant control, e.g., with the usage of model predictive control (MPC), can lead to an increase in plant efficiency. However, the physical modeling of non-linear prediction models used here is often time-consuming and too complex to be used on-line in real-time operation. Instead of physical modeling, though, non-linear prediction models can also be modeled using a fully data-driven approach. An artificial neural network (ANN) used here only requires historical plant data and can therefore learn the non-linear plant behavior itself.

However, data-driven models generally show weaknesses in their extrapolation capabilities, i.e., in areas where no data was available for model training. In this work, a digital twin is therefore deployed to act as an additional data source. By using the digital twin, data-poor operating areas of the plant can be augmented with artificially generated data. The prediction model is then trained with data from both data sources, i.e., historical plant data and data produced with the digital twin. One requirement is, though, that the digital twin must be modeled in close accordance to the plant. Another important issue to address is the plant-model mismatch occurring due to measurement errors at the plant.

The aim of this work is to determine the prediction quality of a hybrid prediction model, trained from different data sources, and compare it with a fully data-driven prediction model. Especially in the data-poor areas, it is expected to achieve a control model with an increased prediction quality. In addition, the uncertainty behavior of the control model is investigated to determine whether this is improved by augmenting the original plant data set.

The hybrid model can be particularly advantageous for newly constructed plants as the time required to acquire a meaningful data set for training the data-driven prediction model can be drastically reduced. This would allow the plant to be operated at high performance more quickly.

617a Keynote Talk: Nonlinear Control for an Industrial Process with Steady-State Gain Inversion

Rahul Bindlish

1. NMPC Methodology

Nonlinear Model Predictive Control (NMPC) is used infrequently for industrial applications compared to Linear Model Predictive Control (LMPC) [1] due to associated modeling and computational burden, and most of these applications are in the field of polymerization control. Industrial use of NMPC is typically justified for processes where LMPC will not be able to achieve process control objectives due to severe nonlinearities in steady-state gains, dead times and dynamic response [2]. A robust NMPC formulation is required to control the output variable at the peak for an industrial application with steady-state gain inversion due to process disturbances and potential modeling errors. The measured feed-forward disturbances along with any unmeasured disturbances also have an effect on the location of the peak or optimum for the controlled variable. Researchers have attempted to address the problem of controlling a process that exhibits change in the sign of steady-state gain for output variable with respect to the manipulated input variable. Lack of stability for an unconstrained nonlinear controller at the peak for processes with steady-state gain inversion has been discussed [3, 4]. The proposed nonlinear model predictive controller augmented with a custom output measurement along with an appropriate disturbance model [5] is different and better because it solves the nonlinear constrained optimization problem explicitly without formulating an unconstrained control law thereby ensuring robustness for controlling at the optimum point where steady-state gain changes sign. Previous simulation work focused on unconstrained controller formulation and appropriate tuning to prevent instability in presence of disturbances for a process with steady-state gain inversion. For ill-conditioned processes with steady-state gain inversion and the control objective of maximizing the output at the peak, the closed-loop controller will exhibit instability when the sign of the gain is different between the model and the plant [6]. In this work, the controller objective is to control at the peak where steady-state gain inversion happens in presence of disturbances instead of preventing instability that arises from operating at that point. Controlling at the optimum in presence of measured and unmeasured disturbances will lead to frequent steady-state gain inversion and needs appropriate estimation of unmeasured plant-model mismatch for proper control action that maximizes the controlled output. Problems arise in applications to control actual industrial processes with gain inversion due to significant process disturbances and potential modeling errors thereby increasing the importance of a robust solution. The novel features of the proposed NMPC to maximize an output variable that has steady-state gain inversion with respect to the manipulated input are

◦ Use of the optimum steady-state manipulated input as a custom output measurement that is available infrequently,

◦ Input disturbance model that utilizes the infrequent custom output measurement to act as an integrator for effective offset free control

The optimum steady-state manipulated input is used as an infrequent custom measurement instead of the available output directly to provide a robust estimate of the input disturbance that accounts for plant-model mismatch and unmodeled disturbances. Robust identification of the input disturbance is important to ensure efficient maximization of the output for noisy industrial data with potential modeling errors.

2. NMPC application

2.1 Process Description

Ethylene oxide is produced by using silver based catalysts for selective oxidation of ethylene to ethylene oxide thereby minimizing secondary reactions that decrease ethylene oxide (EO) selectivity. For conventional catalysts, EO selectivity does not reach values above 85.7 percent that had long been considered as the theoretical maximum selectivity for the process [7]. Modern industrial ethylene epoxidation reactors use co-fed chlorination promoters that are adsorbed on the catalyst to provide moderate chlorine coverage thereby further increasing selective oxidation to ethylene oxide. These high efficiency industrial catalysts also tend to exhibit relatively steep parabolic curves for EO selectivity as a function of adsorbed chloride that is measured as a dimensionless chlorination parameter (Z). The location of the peak or optimum EO selectivity is also a strong function of reaction temperature that is used to control EO production rate. The objective of the feedback controller is to maximize EO selectivity by manipulating the chlorination parameter (Z) in presence of disturbances due to EO production rate and inlet oxygen concentration. The location of peak where the steady-state gain for EO selectivity with respect to chlorination parameter reaches a maximum and changes sign is dependent on measured disturbances and unmeasured disturbances. Unmeasured disturbances that also include the mismatch in predicted catalyst performance as per its age will cause the location of the optimized chlorination parameter for maximizing EO selectivity to be different.

2.2 NMPC Results

EO selectivity has been maximized more effectively using NMPC than the old control scheme that used the steady-state non-linear process model to calculate open-loop targets for chlorination parameter (Z) resulting in an average gain of 0.5 − 1 percent in selectivity for the industrial process after taking catalyst aging and EO production rate effects into account. Three different scenarios have been shown with plant data for the NMPC application to maximize EO selectivity [5]

- EO Selectivity control at high EO production rate
- EO Selectivity control with increasing EO production rate
- EO Selectivity control for a big unmeasured disturbance

The moves for the chlorination parameter, Z, are implemented using the proposed NMPC controller by passing them as targets to the secondary LMPC or PID controller that controls the chlorination parameter by manipulating ethyl chloride flow and rejects faster disturbances. Strong nonlinearities due to gain inversion for maximization of EO selectivity along with noisy measurements for industrial data necessitate the formulation of a novel robust NMPC controller. The computational burden for NMPC has been reduced to enable real-time industrial application by formulating a hierarchical controller for EO selectivity and chlorination parameter.

3. Conclusions

The proposed NMPC controller formulation for a process with steady-state gain inversion utilizes the optimum steady-state manipulated input as a custom infrequent output measurement. The location of optimum steady-state manipulated input where controlled output is maximized depends on both measured and unmeasured disturbances, and is used to update the unmeasured input disturbance estimate infrequently. Accurate identification of the input disturbance is problematic with using the controlled output directly due to associated input multiplicity. Robust identification of input multiplicity is important for a controller that has the goal of staying at the peak with noisy industrial data as measurements. The NMPC controller can also be used for processes that exhibit steady-state gain inversion at a minimum instead of a maximum with the control objective of minimizing at the trough. For a process that may have multiple gain inversions, it will be important to locate the global optimum to maximize the output for the associated non-convex optimization problem. The NMPC controller has been applied successfully with results from industrial ethylene epoxidation reactor that show robustness in maximizing EO selectivity of the effluent ethylene oxide at the peak where steady-state gain inversion occurs [5]. The industrial NMPC has been in use since October 2018 to maximize EO selectivity in presence of measured and unmeasured disturbances that affect the location of the peak where steady-state gain inversion occurs. Due to the success achieved in maximizing EO selectivity that resulted in significant commercial value associated with an average gain of 0.5 − 1 percent in EO selectivity for similar catalyst age and EO production rates, the business unit has made it a part of their strategy to deploy these NMPC applications at other EO production sites.

References

1. J. Qin, T. A. Badgwell, A survey of industrial model predictive technology, Control Engineering Practice 11 (2003) 733–764.
2. Bindlish, Nonlinear model predictive control of an industrial polymerization process, Computers and Chemical Engineering 73 (2015) 43–48.
3. T. Biegler, J. B. Rawlings, Optimization approaches to nonlinear model predictive control, Fourth International Conference on Chemical Process Control, Padre Island, TX (1991) 543–571.
4. Daoutidis, C. Kravaris, Dynamic output feedback control of minimum-phase nonlinear processes, Chemical Engineering Science 47 (7) (1992) 837–849.
5. Bindlish, Nonlinear model predictive control of an industrial process with steady-state gain inversion, Computers and Chemical Engineering 135 (2020) 106739.
6. G. Pannocchia, J. B. Rawlings, Disturbance models for offset-free model-predictive control, AIChE Journal 49 (2) (2003) 426–437.
7. L. Zhang, A. C. Liu, M. Habenschuss, Method of achieving and maintaining a specified alkylene oxide production parameter with a high efficiency catalyst, patent No. US 8,362,284 B2 (January 29 2013).

617b Aveva Process Simulation-Oriented Thermodynamic Property Calculation for a Python Hydrodesulfurization Model

Fernanda Andrade, Antônio Tavernard, Fernando Lima, Márcio Martins, Heleno Bispo

The global shift in crude oil quality, driven by an increase in heavy oil production, has required refineries to process denser feedstocks. These new feedstocks, characterized by high viscosities, significant impurities, and low hydrogen-to-carbon ratios, result in diminished distillate production, thus complicating the refining process [1-4]. Moreover, with the rising concerns about climate change, the industrial sector is expanding its production portfolio. The aim in such expansion is to enhance product quality and optimize operational efficiency taking into account the mitigation of pollutant emissions, such as sulfur [5]. In this context, the quest for solutions that reduce environmental impact and boost fossil energy source efficiency is gaining momentum. Consequently, the adoption of cleaner technologies and innovative, well-established processes like catalytic hydrotreating (HDT) is becoming crucial. These advancements aim to achieve more profitable, sustainable, and socially responsible practices [3,6-8].In light of these developments and challenges in the refining industry, our objective in this work is to leverage the integration between the Python programming environment and the Aveva Process Simulation (APS) software to develop a customized computer model for simulating the HDT process, with a focus on HDS.

Catalytic hydrotreating plays a vital role in enhancing the quality of various oil fractions, ranging from naphtha to vacuum residues. It serves as both a preliminary stage for conversion processes and the final stage in producing fuels that comply with stringent environmental standards, such as gasoline and diesel [3]. This process involves subjecting oil fractions to hydrogenation reactions in the presence of a heterogeneous catalyst, typically composed of metals like molybdenum and nickel. Among these reactions, hydrodesulfurization (HDS) is notable. In HDS, sulfur from organosulfur molecules is captured by the catalyst and reacts with hydrogen gas to form hydrogen sulfide (H2S) [6]. Effective operation of the HDT process requires adjusting and optimizing operating conditions based on feed stream properties and desired product characteristics, ensuring efficiency, safety, and economic viability [3,9,10]. However, the process's complexity often necessitates study and sizing using process simulators, which may not always meet industry demands, leading to gaps in development and optimization [3]. Despite available HDT models, their inflexibility to adapt to specific conditions, such as feedstock variations, compromises simulation accuracy and effectiveness.

In light of this, our work aims to leverage the integration between the Python programming environment and the Aveva Process Simulation (APS) software for simulating the HDT process at a pseudo-steady state, with a focus on HDS, as proposed in the literature [7, 3]. This approach enables the development of a customized and adaptable computer model, where Python is used to apply the mathematical model of a trickle-bed reactor and APS is employed to calculate the necessary thermodynamic properties and physicochemical parameters. With this, the result is an optimized HDS model capable of representing the specific conditions of the process, providing greater control over variables and parameters and enabling a more accurate representation of the involved reaction rates, mass transfers, and operating conditions.

Moreover, the use of APS provides the essential level of thermodynamic precision for simulation, enhancing the robustness of the model developed in Python. Consequently, the integration of the Python programming environment with APS software proves highly beneficial for modeling and simulating complex processes, such as HDS. Emphasizing the ease of implementing this methodology, the rapid integration between the platforms resulted in an effective and practical solution for detailed process analysis.

References

[1] Bai, Y.; Lian, Y.; Zhao, J.; Cao, Z.; Sun, J.; Zhang, H. Thermal-Insulation and Temperature-Resistant Foamed Gel for Thermal Management of Heavy Oil Steam Flooding. *J Mol Liq* **2022**, *359*, 119304, doi:10.1016/j.molliq.2022.119304.

[2] Guan, D.; Chen, Z.; Chen, X.; Zhang, Y.; Qi, Q.; Shi, Q.; Zhao, S.; Xu, C.; Zhang, L. Molecular-Level Heavy Petroleum Hydrotreating Modeling and Comparison with High-Resolution Mass Spectrometry. *Fuel* **2021**, *297*, 120792, doi:10.1016/j.fuel.2021.120792.

[3] Pereira Neto, A.T.; Fernandes, T.C.R.L.; da Silva Junior, H.B.; de Araújo, A.C.B.; Alves, J.J.N. Three-Phase Trickle-Bed Reactor Model for Industrial Hydrotreating Processes: CFD and Experimental Verification. *Fuel Processing Technology* **2020**, *208*, 106496, doi:10.1016/j.fuproc.2020.106496.

[4] Saleh, T.A. Characterization, Determination and Elimination Technologies for Sulfur from Petroleum: Toward Cleaner Fuel and a Safe Environment. *Trends in Environmental Analytical Chemistry* **2020**, *25*, e00080, doi:10.1016/j.teac.2020.e00080.

[5] Chehadeh, D.; Ma, X.; Al Bazzaz, H. Recent Progress in Hydrotreating Kinetics and Modeling of Heavy Oil and Residue: A Review. *Fuel* **2023**, *334*, 126404, doi:10.1016/j.fuel.2022.126404.

[6] Ancheyta, J. *Modeling and Simulation of Catalytic Reactors for Petroleum Refining*; John Wiley & Sons, Inc.: Hoboken, New Jersey, 2011;

[7] Korsten, H.; Hoffmann, U. Three-phase Reactor Model for Hydrotreating in Pilot Trickle-bed Reactors. *AIChE Journal* **1996**, *42*, 1350–1360, doi:10.1002/aic.690420515.

[8] Leal, E.; Morales-Leal, F.; Ancheyta, J.; Alonso, F.; Torres-Mancera, P. Maximizing Data Analysis from Batch Reactor Experiments: Application to Catalytic Hydrotreating of Petroleum Distillates. *Ind Eng Chem Res* **2023**, *62*, 17631–17645, doi:10.1021/acs.iecr.3c02667.

[9] Ancheyta, J.; Alvarez-Majmutov, A.; Leyva, C. Hydrotreating of Oil Fractions. In *Multiphase Catalytic Reactors*; Wiley, 2016; pp. 295–329.

[10] Chowdhury, R.; Pedernera, E.; Reimert, R. Trickle-bed Reactor Model for Desulfurization and Dearomatization of Diesel. *AIChE Journal* **2002**, *48*, 126–135, doi:10.1002/aic.690480113.

617c Predictive Modeling and Optimal Control of a Particulate Polysilicon Reactor System for Enhanced Solar Cell Manufacturing

Carlos Veloz, Davood Babaei Pourkargar

In 2022, the global photovoltaic industry reached the terawatt era, with a cumulative installed capacity of 1.185 GW [1]. Crystal silicon cells represent over 95% of this capacity [2]. Solar-grade silicon, a high-purity polysilicon, is pivotal in the photovoltaic industry, particularly in manufacturing solar panels. It accounts for approximately 20% of the total solar cell manufacturing cost. Consequently, reducing the production cost of solar-grade silicon is a primary factor in enhancing the solar manufacturing process. Several technologies have been explored based on different intermediate Si gaseous species. However, the Siemens process remains the dominant technology for 90% of global polysilicon production. Despite being a mature technology in the photovoltaic industry, it has significant limitations, including its batch-processing nature and energy inefficiency.

Fluidized-bed reactors (FBR) emerge as a promising technology for solar-grade silicon production, representing a more energy-efficient process with more significant operational benefits than the conventional Siemens process. Their performance has been assessed by various experimental [3-4] and computational studies [5-7]. However, controlling the FBR system is challenging due to the complex gas-solid interactions. Several multiphase gas-solid modeling approaches using computational fluid dynamic tools have been developed to predict and analyze FBR system dynamics. Limited research has been conducted on model-based control strategies for enhancing silicon production in FBR systems. Prior research has primarily employed traditional control methods due to the complexity of the models. In addition, the control approaches have been solely used to regulate mass hold-up and particle size distribution, overlooking the objective of minimizing powder loss to enhance process yield. Advanced control strategies, such as model predictive control (MPC), appear promising to address these challenges.

This work presents a predictive modeling framework for silicon production in FBRs, suitable for real-time optimization and control applications. The proposed model characterizes the particle size distribution of the product and the powder loss. Two different flow regime modeling approaches are considered to describe the silane pyrolysis reaction and to represent the deposition rate contributing to particle growth. A discrete population balance equation is employed to estimate the particle size distribution as a function of the deposition rate. Subsequently, a nonlinear model predictive control is utilized to regulate the system towards the desired operating conditions. Detailed open-loop and closed-loop simulation studies demonstrate the successful integration of nonlinear MPC and the proposed predictive modeling approach.

References:

[1] Yu, Y., Bai, X., Li, S., Shi, J., Wang, L., Xi, F., Ma, W., and Deng, R. (2023). Review of silicon recovery in the photovoltaic industry. Current Opinion in Green and Sustainable Chemistry, page 100870.

[2] Ballif, C., Haug, F.-J., Boccard, M., Verlinden, P. J., and Hahn, G. (2022). Status and perspectives of crystalline silicon photovoltaics in research and industry. Nature Reviews Materials, 7(8):597-616.

[3] Cadoret, L., Reuge, N., Pannala, S., Syamlal, M., Coufort, C., and Caussat, B. (2007). Silicon CVD on powders in fluidized bed: Experimental and multifluid Eulerian modeling study. Surface and

Coatings Technology, 201(22-23):8919-8923.

[4] Hashimoto, K., Miura, K., Masuda, T., Toma, M., Sawai, H., and Kawase, M. (1990). Growth kinetics of polycrystalline silicon from silane by thermal chemical vapor deposition method. Journal of the Electrochemical Society, 137(3):1000.

[5] Chanlaor, P., Limtrakul, S., Vatanatham, T., and Ramachandran, P. A. (2018). Modeling of chemical vapor deposition of silane for silicon production in a spouted bed via discrete element method coupled with Eulerian model. Industrial & Engineering Chemistry Research, 57(36):12096-12112.

[6] Du, J., Dutta, S., and Ydstie, B. E. (2014). Modeling and control of solar-grade silicon production in a fluidized bed reactor. AIChE Journal, 60(5):1740–1751.

[7] Tejero-Ezpeleta, M. P., Buchholz, S., and Mleczko, L. (2004). Optimization of reaction conditions in a fluidized bed for silane pyrolysis. The Canadian Journal of Chemical Engineering, 82(3):520–529.

617d Dynamic Risk-Based Batch Reactor Quality Control through Hierarchical Multi-Parametric Optimization

Austin Braniff, Beatriz Dantas, Sahithi Srijana Akundi, Yuanxing Liu, Shayan Niknezhad, Faisal Khan, Efstratios Pistikopoulos, Yuhe Tian

Batch reaction processes are a cornerstone in the chemical manufacturing industry widely applied to produce pharmaceuticals, high-value added chemicals, etc. However, their inherently dynamic operations have been a central question to control engineers as the product quality remains unknown until the end of a batch [1]. Extensive research works have been developed for batch quality control, such as the integration of principal component analysis with statistical process control or model predictive control (MPC) [2-4]. Several recent works also explored the use of nonlinear MPC through Gaussian process techniques and robust constraints with application in semi-batch processes [5-7]. As the decision making of regulatory process control and end-batch quality control span multiple time scales, it typically results in a large-scale dynamic optimization problem which poses challenges in the online computational efficiency and algorithmic tractability. In addition, the paradigm shift toward Industry 4.0 has emphasized the importance of proactive process safety management through advanced control [8-10]. A multi-time-scale operational optimization approach is thus essential, while currently lacking, to systematically identify the safe and optimal operating trajectories by accounting for the interactions and trade-offs between process control, prescriptive risk management, and end-batch quality control.

To address this gap, we present a methodology framework for dynamic risk-based model predictive quality control, as an extension of our recent work on risk-based control in continuous processes [11]. A statistical dynamic risk model is utilized as the indicator of real-time process safety performance which is a function of safety-critical process variables (e.g., temperature, pressure) [12]. The batch reactor model can adapt a first-principle model, data-driven model using measured data, or hybrid physics-informed model updating [13]. On this basis, the proposed framework features three hierarchical decision-making algorithms: (i) Short-term risk-aware controller – An MPC problem is formulated based on the integrated dynamic process and risk model to operate the batch systems with desired process safety performance. The MPC moving horizon estimation allows to forecast the process risk into the near future, thereby raising alarms ahead of fault occurrence when necessary. (ii) Long-term quality and safety optimizer – A batch quality model is developed utilizing information from the closed-loop risk-aware control. In this way, the batch quality of the system can be mathematically mapped as a function of process state variables and the setpoints of the risk-aware controller. The batch quality model must be on a larger time scale to be able to predict the end-point quality of the reactor. (iii) Intermediate surrogate model – For substantially different time scales, the batch quality model can omit crucial intermediate dynamics of the system or suggest overly aggressive changes in set points. In such cases, a surrogate model is necessitated to interpret the input and output setpoints from the optimizer for the risk-aware controller over an intermediate time scale to ensure smooth operational transitions. In this way, the decisions from upper-level decision making (in longer term) can be seamlessly conveyed to the lower-level decision making (in shorter term), while the lower-level dynamics are explicitly incorporated into the long-term representation. Notably, all of these three decision making problems are formulated as multi-parametric (mixed-integer) quadratic programming problems, thereby maximally replacing online optimization with explicit solutions and optimal look-up maps generated a priori via offline optimization. The efficacy of this framework is demonstrated through a case study on a safety-critical batch reactor that is conceptualized from the T2 Laboratories incident.

References

1. [1] Mhaskar, P., Garg, A., & Corbett, B. (2019). Modeling and Control of Batch Processes: Theory and Applications. Springer International Publishing.
2. [2] Nomikos, P., & MacGregor, J. F. (1994). Monitoring batch processes using multiway principal component analysis. AIChE Journal, 40(8), 1361–1375.
3. [3] Nomikos, P., & MacGregor, J. F. (1995). Multi-way partial least squares in monitoring batch processes. Chemometrics and Intelligent Laboratory Systems, 30(1), 97–108.
4. [4] Aumi, S., Corbett, B., & Mhaskar, P. (2012). Model predictive quality control of batch processes. 2012 American Control Conference (ACC), 5646–5651.
5. [5] Jang, H., Lee, J. H., & Biegler, L. T. (2016). A robust NMPC scheme for semi-batch polymerization reactors. IFAC-PapersOnLine, 49(7), 37–42.
6. [6] Chang, L., Liu, X., & Henson, M. A. (2016). Nonlinear model predictive control of fed-batch fermentations using dynamic flux balance models. Journal of Process Control, 42, 137–149.
7. [7] Thombre, M., (Joyce) Yu, Z., Jäschke, J., & Biegler, L. T. (2021). Sensitivity-Assisted multistage nonlinear model predictive control: Robustness, stability and computational efficiency. Computers & Chemical Engineering, 148, 107269.
8. [8] Khan, F., Amyotte, P., & Adedigba, S. (2021). Process safety concerns in process system digitalization. Education for Chemical Engineers, 34, 33–46.
9. [9] Bhadriraju, B., Kwon, J. S. I., & Khan, F. (2021). Risk-based fault prediction of chemical processes using operable adaptive sparse identification of systems (OASIS). Computers & Chemical Engineering, 152, 107378.
10. [10] Khan, F., Hashemi, S. J., Paltrinieri, N., Amyotte, P., Cozzani, V., & Reniers, G. (2016). Dynamic risk management: A contemporary approach to process safety management. Current Opinion in Chemical Engineering, 14, 9–17.
11. [11] Ali, M., Cai, X., Khan, F. I., Pistikopoulos, E. N., & Tian, Y. (2023). Dynamic risk-based process design and operational optimization via multi-parametric programming. Digital Chemical Engineering, 7, 100096
12. [12] Bao, H., Khan, F., Iqbal, T., Chang, Y. (2011). Risk-based fault diagnosis and safety management for process systems. Process Saf. Prog., 30(1), 6-17.
13. [13] Wang, W., Wang, Y., Tian, Y., & Wu, Z. Explicit Machine Learning-Based Model Predictive Control of Nonlinear Processes via Multi-Parametric Programming.

617e Enhancing Poly(3HB-*co*-4HB) Production: Incorporating Multiscale Model into MPC to Control Molecular Weight Distributions and Monomeric Ratios

Juhyeon Kim, Parth Shah, Raj Panthesh Bhavsar, Joseph Kwon

While petroleum-derived plastic materials have brought significant convenience into the daily life of mankind, they are predominantly non-renewable and unsustainable. Despite their easy production, their disposal poses serious environmental risks, including air/water pollution, greenhouse gas emissions, and the release of toxic substances [1], highlighting the urgent need for eco-friendly polymer alternatives. To overcome this issue, the exploration of bio-based polymer production, particularly, polyhydroxyalkanoates (PHAs), has attracted research interests due to their biocompatibility, synthesis from natural resources, and marine degradability [2].

Among diverse PHAs, the 3-hydroxybutyrate (3HB) homopolymer is well recognized and utilized. Nevertheless, its brittleness and stiffness limit its usability, necessitating the development of more adaptable varieties [3]. In this sense, some PHA variants such as poly(3-hydroxybutyrate-*co*-4-hydroxybutyrate) (P(3HB-*co*-4HB)) have been developed [4]. It can offer more flexibility and its properties can be tuned by controlling their monomeric composition. However, there are many factors determining their physical properties. To address this, a multiscale kMC model was developed to elucidate the intricate polymerization dynamics of P(3HB-*co*-4HB) [5].

As biopolymer production involves cell metabolism and polymerization, the overall process was modeled with the two-layer structure, handling significant difference in their length/time scales. The first layer was modeled by a series of differential equations coupled with Monod equation. This is categorized as macroscopic reactions, including the carbon source consumption, monomer production, by-product generation, as functions of pH, feed rate, temperature [6]. Produced biomass (*e.g.*, synthase) and monomers are introduced in the microscopic kMC layer as reacting particles. In each time step, the rates of initiation, elongation, termination, and depolymerization were evaluated. Subsequently, the particles interact and form specific molecular weight distribution (MWD) and monomeric composition. This multiscale model was not only validated against the pilot-scale lab experiments, but also shown the outstanding capabilities of capturing MWDs and their monomeric compositions.

Subsequently, the multiscale model is seamlessly incorporated into the model predictive control (MPC) framework. This novel approach achieves unprecedented control performance with delicate objectives – including precise regulation of MWD and monomeric ratios – surpassing the limitations of average-value controls. This result highlights the effective coupling of the multiscale model with MPC, taking advantage of the strengths of both strategies to offer a comprehensive control solution. In conclusion, this work will shed light on developing an advanced control strategy in complex, multiscale reaction systems. Also, it provides substantial enhancements in efficiency and sustainability in biopolymer production to reduce environmental impacts markedly.

Literature cited:

[1] Ewing T.A., Nouse N., van Lint M., van Haveren J., Hugenholtz J., & van Es D.S. (2022). Fermentation for the production of biobased chemicals in a circular economy: a perspective for the period 2022-2050. *Green Chem.*, 24, 6373-6405.

[2] Naser A.Z., Deiab I., & Darras B.M. (2021). Poly(lactic acid) (PLA) and polydydroxyalkanoates

(PHAs), green alternatives to petroleum-based plastics: a review. *RSC Adv.*, 11, 17151-17196.

[3] Chotchindakun K., Pathom-Aree W., Dumri K., Ruangsuriya J., Pumas C., & Pekkoh J., (2021). Low crystallinity of poly(3-hydroxybutyrate-*co*-3-hydroxyvalerate) bioproduction by hot spring cyanobacterium *Cyanosarcina* sp. AARL T020. *Plants*, 10, 503.

[4] Huong K.H., Sevakumaran V., & Amirul A.A., (2021). P(3HB-co-4HB) as high value polyhydroxyalkanoate: its development over recent decades and current advances. *Crit. Rev. Biotechnol.*, 41, 474-490.

[5] Kim, J., Pahari, S., Shah, P., & Kwon, J.S.-I., (2023). Enhanced Polyhydroxyalkanoate (PHA) Production through Multiscale Modeling and Process Control Strategies: A Novel Approach to Bio-Based Polymer Synthesis. In *2023 AIChE Annual Meeting*. AIChE.

[6] McAdam B., Fournet M.B., McDonald P., & Mojicevic M., (2020). Production of Polyhydroxybutyrate (PHB) and factors impacting its chemical and mechanical characteristics. *Polymers*, 12, 2908.

617f Dynamic Embedded Gaussian Process Model Applied to Steam Methane Reforming

Claudemi Nascimento, Michael Fouts, David S. Mebane, Fernando Lima

Uncertainties from reduced-order models derived from small-scale systems can impact the accuracy and reliability of predictions for large-scale systems [1]. Recent developments in machine learning (ML) techniques have opened avenues to propose new ways to build fast and accurate scalable models. For this purpose, a model selection framework for the generation of Gaussian Process (GPs) models with a Bayesian Smoothing Spline Analysis of Variance (BSS-ANOVA) using Karhunen-Loève (KL) decomposition has been proposed, known as FoKL-GPs [2]. This approach has been employed to model different scalable systems focusing on applications including kinetics, thermodynamics, and process control [1-5]. In this work, FoKL-GPs are employed to construct novel, accurate, and scalable dynamic embedded GPs, focusing on the case study of a Steam Methane Reformer (SMR). By combining a classical SMR reduced-order model and an ab initio (first-principles) model in a grey-box fashion with FoKL-GPs, a more accurate and scalable model can be achieved for online optimization and control.

The embedded GP model accounts for the mismatch between the classical kinetic SMR model proposed by Xu and Froment [6] using Langmuir equilibrium relations to eliminate adsorbed species and an ab initio reaction scheme derived from microkinetic modeling [7]. The ab initio microkinetic model implemented in this work is based on thermochemical data obtained from Density Functional Theory (DFT) and statistical thermodynamics for nickel-catalyzed steam methane reforming on Ni (111) under industrial reforming conditions (800 °C, 10 bar) [8]. This microkinetic model for the SMR reaction brings relevant information about the surface in the catalytic process, which allows the assessment of which adsorbed species have relevant or nonrelevant concentrations on the surface and how they influence concentration profiles of the gas-phase species leaving the system.

FoKL-GPs calibrate the mismatches between the ab initio and classical SMR models using a series of dynamic embedded Gaussian Process models. The calibrated models are then further incorporated into the rate of change equations of the classical model, and validation followed by upscaling are performed by comparing the concentration profiles of the embedded models in different scales against the classical and the first-principles models. These obtained models provide fast predictions for use in subsequent dynamic real-time optimization and advanced process control. Combining first-principles information within reduced-order models can accelerate the development of new technologies and algorithms in process systems engineering using accurate and reliable predictions.

References

[1] Bhat, K.S., Mebane, D.S., Mahapatra, P., Storlie, C.B., 2017. Upscaling Uncertainty with Dynamic Discrepancy for a Multi-Scale Carbon Capture System. Journal of the American Statistical Association 112, 1453–1467.

[2] Hayes, K., Fouts, M.W., Baheri, A., Mebane, D.S., 2022. Forward variable selection enables fast and accurate dynamic system identification with Karhunen-Loève decomposed Gaussian processes. e-print arXiv: 2205.13676.

[3] Ostace, A., Chen, Y.-Y., Parker, R., Mebane, D.S., Okoli, C.O., Lee, A., Tong, A., Fan, L.-S., Biegler, L.T., Burgard, A.P., Miller, D.C., Bhattacharyya, D., 2022. Kinetic model development and

Bayesian uncertainty quantification for the complete reduction of Fe-based oxygen carriers with CH_4, CO, and H_2 for chemical looping combustion. Chemical Engineering Science 252, 117512.

[4] Mebane, D.S., Bhat, K.S., Kress, J.D., Fauth, D.J., Gray, M.L., Lee, A., Miller, D.C., 2013. Bayesian calibration of thermodynamic models for the uptake of CO_2 in supported amine sorbents using ab initio priors. Phys. Chem. Chem. Phys. 15, 4355.

[5] Dinh, S., Mebane, D.S., Lima, F.V., 2023. A Framework for Implementation of Dynamic Discrepancy Reduced-Order Modeling in Advanced Process Control. Submitted for Publication.

[6] Xu, J., Froment, G.F., 1989. Methane steam reforming, methanation and water-gas shift: I. Intrinsic kinetics. AIChE J. 35, 88–96.

[7] Motagamwala, A.H., Dumesic, J.A., 2021. Microkinetic Modeling: A Tool for Rational Catalyst Design. Chem. Rev. 121, 1049–1076.

[8] Blaylock, D.W., Ogura, T., Green, W.H., Beran, G.J.O., 2009. Computational Investigation of Thermochemistry and Kinetics of Steam Methane Reforming on Ni(111) under Realistic Conditions. J. Phys. Chem. C 113, 4898–4908. https://doi.org/10.1021/jp806527q

564b Investigating the Dynamic Response of a Process to Control Implemented on a Quantum Computer

Shilpa Narasimhan, Henrique Oyama, Dominic Messina, Helen Durand

Traditional process control systems are classical computers which receive feedback from a chemical manufacturing process and utilize a control law to compute the control input that is implemented on the process to operate it in a safe and economical manner. As a result, advances in computing systems are important to analyze within a control context, to see whether they may hold benefits. One advance in computing that is receiving some attention within the process systems engineering community is quantum computation [1-5]. Quantum computing is a computing methodology based on quantum principles, and as a result quantum devices operate on fundamentally different principles than classical computers (and thus can run fundamentally different algorithms).

A key challenge for quantum computation today is that the internal states of a quantum computer are susceptible to noise within quantum circuit, and therefore may introduce nondeterminism (errors) in the computations performed by a quantum computer [6]. An additional source of nondeterminism may be the quantum algorithm itself. As a result of both of these sources of non-determinism, it is unclear how applicable quantum algorithms may be within a process control context. The accuracy of the control input computed by a process control system is a crucial requirement to ensure the stability and profitable operation of a chemical manufacturing process. Consequently, handling nondeterminism within quantum circuits is an important consideration for the successful implementation of process control laws on a quantum computer. Motivated by this, prior work by our group considered the analysis and design of control algorithms that may be implemented on a quantum computer [3-5]. In [4], an analysis was presented which considered the application of a quantum algorithm (based on Grover's algorithm [6]) that searches for optimal paths over a graph as a means to implement an MPC on an illustrative chemical process example. The proposed work considered control implemented on a noise-free quantum computer. In [5], an analysis considering if the nondeterminism within a quantum circuit may fortify a process control system against cyberattacks was presented. The analysis revealed that nondeterminism within a quantum circuit does not, on its own, protect a system from cyberattacks on the process control system. Finally, in [3], the stability of an illustrative process example in the presence of different levels of errors simulating the noise within quantum circuits was analyzed. Our prior analyses have focused primarily on pathfinding examples (e.g., a process example with a single state and a single control input with a linear control law computed using a deterministic quantum algorithm). However, chemical manufacturing processes have complex process dynamics and the dynamic response of the illustrative process example to nondeterminism may not be indicative of the dynamic response of a real-time chemical process.

In this work, we seek to better understand the impacts of noise and non-determinism on control implemented with quantum devices on chemical processes. In particular, we consider several key practical issues: 1) the significance of the noise (i.e., tolerable levels of noise for meeting production targets, and how the noise interacts with other process disturbances); 2) the role of time constants in, for example, sensors in damping out the noise; and 3) the effects of spatial-temporal phenomena in damping out the noise. To analyze these points, we will consider proportional-integral-derivative (PID)-type control laws implemented with at least one of the additions in the computations implemented with a quantum simulator with noise [8]. We will consider this control law for a benchmark continuous stirred tank reactor (CSTR) with sensor noise and process disturbances, and analyze how the added noise from the quantum device impacts the closed-loop performance. Next, we will consider a wafer temperature control (WTC) system supporting the plasma etching process for the semiconductor industry [7]. A model that we developed for this process in [7] contains the information about the dynamics of the actuators and the sensors, and therefore can be used for analyzing the role of the quantum noise (and the sampling period/input selection frequency) on sensor readings. In particular, we analyze the conditions under which the fluctuations due to the quantum noise are smoothed out, and when the quantum noise effects become more prominent. Finally, we will discuss how distributed parameter systems, simulated in ANSYS Fluent, respond to the noise in the device. For this, we will utilize a benchmark simulation of flow in a pipe with a sensor simulated within the pipe [9]. We will develop a workflow for simulating quantum algorithms with noise though user-defined functions (UDF's) in Fluent, and will then analyze the extent to which both the spatio-temporal dynamics, as well as sensor dynamics, cause the impacts of the noise from the controller to be damped out or amplified.

References:

[1] A., Ajagekar, and F., You, "New frontiers of quantum computing in chemical engineering", Korean Journal of Chemical Engineering, volume 39, pp. 811–820, 2022.

[2] Nourbakhsh A, Jones M.N., Kristjuhan K., Carberry D., Karon J., Beenfeldt C., Shahriari K., Andersson M. P., Jadidi M. A., Mansouri S.S., "Quantum computing: Fundamentals, trends and perspectives for chemical and biochemical engineers", arXiv preprint arXiv:2201.02823, 2022.

[3] Nieman, K., Rangan, K. K., and Durand, H., "Control Implemented on Quantum Computers: Effects of Noise, Non-Determinism, and Entanglement", Industrial & Engineering Chemistry Research, volume 61(28), pp. 10133-10155, 2022.

[4] Nieman, K., Durand, H., Patel, S., Koch, D. and Alsing, P.M., "Investigating an amplitude amplification-based optimization algorithm for model predictive control", Digital Chemical Engineering, volume 10, p.100134, 2024.

[5] Rangan, K. K., Abou Halloun, J., Oyama, H., Cherney, S., Assoumani, I. A., Jairazbhoy, N., Durand, H, and Ng, S.K., "Quantum computing and resilient design perspectives for cybersecurity of feedback systems", Proceedings of the 13th IFAC Symposium on Dynamics and Control of Process Systems, including Biosystems, volume 55(7), pp. 703-708, Busan, Republic of Korea, 14–17 June 2022.

[6] Córcoles, A.D., Kandala, A., Javadi-Abhari, A., McClure, D.T., Cross, A.W., Temme, K., Nation, P.D., Steffen, M., Gambetta, J.M., "Challenges and opportunities of near-term quantum computing systems" Proceedings of the IEEE, volume 108(8), pp. 1338-1352, December 2019.

[7] Oyama, H., Nieman, K., Tran, A., Keville, B., Wu, Y., Durand, H., "Computational fluid dynamics modeling of a wafer etch temperature control system", Digital Chemical Engineering, volume 8, p. 100102, 2023.

[8] Norlén H., "Quantum Computing in Practice with Qiskit® and IBM Quantum Experience®: Practical recipes for quantum computer coding at the gate and algorithm level with Python", Packt Publishing Ltd, 2020.

[9] Rangan, Keshav Kasturi, Henrique Oyama, Ilham Azali Assoumani, Helen Durand, and K. Y. Simon Ng. 'Cyberphysical Systems and Energy: A Discussion with Reference to an Enhanced Geothermal Process.' In *Energy Systems and Processes: Recent Advances in Design and Control*, pp. 8-1. Melville, New York: AIP Publishing LLC, 2023.

564g Life Cycle Assessment of Ti_3C_2 MXene: Sustainable Manufacturing Pathways for Chemical and Energy Systems

Mostafa Dadashi Firouzjaei, Mark Elliott, Babak Anasori

Ti_3C_2 MXene, a two-dimensional transition metal carbide, is a promising material for next-generation chemical and energy systems, with applications in energy storage, water purification, and electromagnetic shielding. This study performs a detailed Life Cycle Assessment (LCA) of Ti_3C_2 to evaluate its environmental impacts across the production lifecycle, utilizing the ecoinvent 3.9.1 database and OpenLCA software. Critical stages, including titanium and aluminum extraction, MAX phase synthesis, HF and HCl etching, and post-etching treatments, were examined, with energy-intensive processes such as high-temperature synthesis and centrifugation identified as key contributors to the environmental footprint. To advance sustainable manufacturing, strategies such as energy optimization, the use of safer etching alternatives, and recycling of chemicals are proposed. By comparing the environmental performance of Ti_3C_2 to other advanced materials, this study offers a framework for integrating sustainability into the production of Ti_3C_2 MXene, providing insights for researchers, industries, and policymakers to guide responsible manufacturing practices in chemical and energy applications.

564d Predictive Modeling and Optimal Regulation of Modular Hydrogen-Ammonia Systems with Renewable Energy Integration for Resilient Chemical-Energy Conversion

Thiago Oliveira Cabral, Davood Babaei Pourkargar

The drive to implement more sustainable chemical systems in the 21st century has motivated and propelled the scientific community to investigate innovative, multi-functional processes [1]. Numerous studies have explored the feasibility and resilience of integrating chemical systems and renewable energy technologies, focusing on four potential economic entities: solar and wind power, hydrogen, and ammonia [2-5]. The former two are low-cost, intermittent, and accessible clean energy sources, while the latter two are commodities with crucial applications across various sectors of a functional economic chemical-energy matrix. Hydrogen is produced in large quantities by catalytic steam and dry reforming of hydrocarbons, a non-renewable process that generates significant atmospheric pollution in CO and CO_2. Recent technological advancements have made it possible to competitively produce hydrogen through greener processes, such as water electrolysis. Conversely, ammonia is industrially produced at high pressures by catalytic hydrogenation of atmospheric N_2 through the Haber-Bosch process, relying on air separation technologies, typically adsorption and hydrogen supply.

Other studies propose a new paradigm for chemical manufacturing, shifting from the few large-scale production complexes existing today towards more compact, locally active production systems [6,7]. The benefits of modularized, self-sufficient, and integrated chemical and renewable energy systems include maximizing economic metrics, physical intensification between processes, and reconciling sustainability with environmental concerns. Naturally, this new perspective involves significant technical challenges, such as inherent operational restrictions, intermittency of renewable energy supplies, highly integrated chemical and energy systems, and real-time automation and control. In this context, computational modeling investigations can provide meaningful insights into the operation of modular chemical-energy systems and the necessary predictive capabilities for informed decision-making and control.

While there is relevant work on modeling advanced chemical processes such as hydrogen and ammonia synthesis with different levels of physical intensification (chemical-energy polygeneration, material recycles, etc.), the majority of these studies have only minimally addressed real-time decision-making and control challenges, particularly when considering the direct involvement of renewable energy inputs [8-10]. This limitation is primarily due to complications in the mathematical formulation and numerical solution of the resulting highly integrated and nonlinear differential and algebraic equations. This work develops a phenomenological dynamic model for a modular renewable energy-driven system that simultaneously produces hydrogen and ammonia for resilient energy storage and fertilizer production in rural communities. Specifically, this study focuses on integrating the chemical production system with renewable energy provisions through concentrated solar-thermal and wind power technologies. The hydrogen-ammonia production module incorporates a dynamic axial-dispersion formulation that considers both kinetics and thermodynamic effects. Additionally, other units, such as compressors and heat exchangers, are integrated into the system model to harmonize the distinct operations of the water electrolyzer and ammonia reactor. The solar plant is modeled using transient mass and energy balances that account for variations in solar irradiance throughout the day. Furthermore, a correlation is utilized to directly relate blade height and wind velocity to the output power, facilitating the prediction of wind turbine dynamics [11]. Furthermore, the model efficiency and the integrated system's stability and performance are investigated using model predictive control under constraints imposed by the availability of renewable energy. The controller determines the optimal amount of additional energy input to be acquired from external sources (such as the grid or non-renewables) to achieve the desired steady state.

References:

[1] Thomas A. Adams. Future opportunities and challenges in the design of new energy conversion systems. Computers and Chemical Engineering 81 (2014).

[2] Prodromos Daoutidis, Michael Zachar, and Sujit S. Jogwar. Sustainability and process control: A survey and perspective. Journal of Process Control 44 (2016).

[3] Matthew J. Palys, Hanchu Wang, Qi Zhang, and Prodromos Daoutidis. Renewable ammonia for sustainable energy and agriculture: vision and systems engineering opportunities. Current Opinion in Chemical Engineering 31 (2021).

[4] Jussi Ikaheimo, Juha Kiviluoma, Robert Weiss, and Hannele Holttinen. Power-to-ammonia in future North European 100% renewable power and heat system. International Journal of Hydrogen Energy 43 (2018).

[5] Claus Hviid Christensen, Tue Johannessen, Rasmus Zink Sørensen, and Jens K Nørskov. Towards an ammonia-mediated hydrogen economy? Catalysis Today 111 (2006).

[6] Yong-ha Kim, Lydia K. Park, Sotira Yiacoumi, and Costas Tsouris. Modular chemical process intensification: a review. Annual review of chemical and biomolecular engineering, 8 (2017).

[7] Robert S, Wegeng, M. Kevin Drost, and David L. Brenchley. Process intensification through miniaturization of chemical and thermal systems in the 21st century. In Microreaction Technology: Industrial Prospects, Springer (2000).

[8] Efstratios N. Pistikopoulos, Yuhe Tian, and Rahul Bindlish. Operability and control in process intensification and modular design: Challenges and opportunities. AIChE Journal, 67 (2021).

[9] Nikola M. Nikacevic, Adrie E.M. Huesman, Paul M.J. Van den Hof, and Andrzej I. Stankiewicz. Opportunities and challenges for process control in process intensification. Chemical Engineering and Processing: Process Intensification 52 (2012).

[10] Lisia S. Dias and Marianthi G. Ierapetritou. Optimal operation and control of intensified processes—challenges and opportunities. Current Opinion in Chemical Engineering 25 (2019).

[11] Andrew Allman and Prodromos Daoutidis. Optimal scheduling for wind-powered ammonia generation: Effects of key design parameters. Chemical Engineering Research and Design, 131:5–15 (2018)

564e Bayesian Optimization of a Continuous Polymer Precipitation Process

Zhiqiang Fan, Qing Pu, Baxton Owen, Neal LaBanca, Walter Barnes

Polymer purification is a crucial process in chemical engineering because impurities can greatly affect polymer properties. In the semiconductor industry, polymer precipitation is a common technique for polymer purification. Compared to batch polymer precipitation, continuous polymer precipitation promises easy scale-up of polymer purification and significant energy and cost savings. We have designed a continuous polymer precipitation system that utilizes static mixers, peristaltic pumps, and a phase separator. In this system, an antisolvent precipitation method is used. A secondary solvent known as antisolvent, or precipitant, is added to the solution, resulting in the reduction of the solubility of the solute in the original solvent and consequently generating a supersaturation driving force. After the polymer chains collapse, aggregate, and the precipitation reaction takes place, the liquid waste is continuously isolated from the solid polymer, and the solid polymer is completely dried through a phase separation unit. The purified polymer is then re-diluted to the same conditions as the original polymer solution.

The development of an effective continuous polymer precipitation process requires careful thought of process parameters. For antisolvent precipitation processes, these include the ratio of antisolvent to polymer solution, the flow rate of polymer solution, and the retention rating of the phase-separation component. This study continues the 2023 AIChE presentation ("An Efficient, Cost-Effective, Continuous Polymer Purification Method"), focusing on a self-optimization machine learning method, i.e., Bayesian optimization (BO), that expedites the optimization of experimental parameters' design space. In this study, BO is used to predict the combination of process parameters that achieve desired outcome measures, including impurity reduction, reduction in polydispersity index (PDI), etc. BO designs a series of experiments to gradually locate the desired optimum. Results obtained demonstrate that the sequence of generated experiments allows pinpointing operating conditions for target outcome measures. The implementation of BO led to reduced experimental effort and increased efficiency in process development. The transition to continuous polymer precipitation combined with the Bayesian optimization technique advances manufacturing process development to the next level.

564f Practical Applications of Generative Artificial Intelligence in Process Simulation

Gustavo Silva, Vitor Santos Sr., Nilton Silva, Rohit Gheyi, Heleno Bispo

Artificial Intelligence (AI) has been consolidating as a powerful, palpable, and widespread tool in many areas. AI is revolutionizing, and potentially surpassing, some human capabilities, sparking both visionary predictions and concerns among leading figures [1-4]. Recently, AI technologies, especially deep learning, have made significant strides in processing existing data and generating new designs, setting them apart from conventional expert systems and biologically inspired algorithms. Generative AI design, which utilizes machine learning algorithms to learn from data and create new content, has been recognized as one of the top ten technology trends for 2023 [5]. Intelligent generative technologies such as, ChatGPT by OpenAI, Gemini by Google [7], Mistral by MistralAI, Claude by Anthropic and LLaMA by Meta AI have demonstrated the versatility of generative AI across various fields, establishing it as a leading, and in some cases, a priority area of research. In this context, this kind of environment can help engineers navigate through many design processes based on architectural plans with multiple constraints such as compliance and cost-effectiveness. Thus, the end-to-end design process of generative AI mirrors that of structural engineering, equipped with robust learning and generative capabilities to address the complexities of intelligent design [8]. This journey, in our work, begins with a foundational prototyping phase, where an initial simulation is created based on the principles of GPT's learning derived from the AVEVA Process Simulation (APS) Scripting, a specific tool that allows direct communication between APS and programming language as Python or C#. This initial step is a test related to examining GPT's cognitive abilities in understanding the complex codes and processes inherent in the APS-Python communication domain. This early stage is of paramount importance, revealing a range of challenges and opportunities while also validating the technical and practical feasibility of the project [2]. The initial results have demonstrated that this integration goes beyond simply demonstrating AI's effectiveness in streamlining process integration, heralding a new era of innovation in process engineering. The integration of AI with advanced simulation offers numerous benefits, including reducing operational costs and improving industrial efficiency while adeptly handling the complexities and variabilities of production environments. The combination of AI methodologies based on GPT will redefine the landscape of process integration and optimization within industrial settings, giving process engineers the possibilities to develop deeper insights and more specific applications where human creativity is needed. In conclusion, the development of tools that facilitate the initial steps of process simulation is essential in the modern engineering landscape. By automating routine tasks and streamlining the integration of AI and simulation platforms, engineers are liberated from mechanical procedural work. This freedom allows them to focus their efforts on more creative and innovative aspects of process design and optimization. Such tools enhance efficiency and productivity resulting in advancements in process engineering, where human and AI-driven insights converge to shape the future of industrial processes.

504a Model-Based Analysis of the Monoclonal Antibody Glycosylation across CHO Cell Lines

Jing Guo, Yingjie Ma, Richard Braatz

N-linked glycosylation, a post-translational modification where the glycan is attached to the nitrogen atom of asparagine in a protein, is a critical quality attribute in biologic manufacturing because it can affect the half-life, immunogenicity, and pharmacokinetics of therapeutic monoclonal antibodies (mAbs) [1]. To maintain a consistent glycosylation profile, multiple mechanistic models have been developed to predict glycosylation profiles for both fed-batch [2] and perfusion [3] cell culture processes. While the mechanisms and associated model are expected to be consistent across cell lines, some of the specific model parameters may differ between cell lines. Despite all Chinese hamster ovary (CHO) cell lines sharing a common ancestor, extensive mutagenesis and clonal selection has resulted in substantial genetic heterogeneity among them [4].

To evaluate the influence of these genetic differences on the bioprocess and mAb quality, we establish a multiscale glycosylation model and adapt the model to various cell lines and culture conditions. Due to the complex glycan measurements and high cost of cell culture experiments, there are limited data available for model evaluation and parameter estimation. Therefore, we employ Bayesian estimation to incorporate prior knowledge from the existing literature and alleviate non-identifiability problems. The uncertainties of model parameters are then propagated to quantify the uncertainties of key model prediction variables. To improve the computational speed of parameter estimation, we employ several strategies including a decomposition method, a simplified glycosylation reaction network, and a tailor-made parallelization scheme.

With the above parameter estimation framework, we obtain and compare the probability distribution for each model parameter among various cell lines. While many parameters maintain similar distribution, some concentration and kinetic parameters of certain Golgi-resident enzymes exhibit significant changes across different cell lines. The expression of these enzymes is likely upregulated or downregulated through genetic engineering methods within cell line development. This quantitative analysis of the host cell impact on mAb glycosylation profiles provides insight into future glycoengineering strategies for improved product quality.

[1] E. Edwards, M. Livanos, A. Krueger, A. Dell, S.M. Haslam, C. Mark Smales, D.G. Bracewell, Strategies to control therapeutic antibody glycosylation during bioprocessing: Synthesis and separation, Biotechnol Bioeng 119(6) (2022) 1343-1358.

[2] P. Kotidis, P. Jedrzejewski, S.N. Sou, C. Sellick, K. Polizzi, I.J. Del Val, C. Kontoravdi, Model-based optimization of antibody galactosylation in CHO cell culture, Biotechnol Bioeng 116(7) (2019) 1612-1626.

[3] D.J. Karst, E. Scibona, E. Serra, J.M. Bielser, J. Souquet, M. Stettler, H. Broly, M. Soos, M. Morbidelli, T.K. Villiger, Modulation and modeling of monoclonal antibody N-linked glycosylation in mammalian cell perfusion reactors, Biotechnol Bioeng 114(9) (2017) 1978-1990.

[4] D. Reinhart, L. Damjanovic, C. Kaisermayer, W. Sommeregger, A. Gili, B. Gasselhuber, A. Castan, P. Mayrhofer, C. Grunwald-Gruber, R. Kunert, Bioprocessing of Recombinant CHO-K1, CHO-DG44, and CHO-S: CHO Expression Hosts Favor Either mAb Production or Biomass Synthesis, Biotechnol J 14(3) (2019) e1700686.

504b Modeling and Performing Multiple Dual Mode Countercurrent Chromatography for Challenging Bioseparations

Hoon Choi, Manar Alherech, Gregg T. Beckham

Valuable targets often require multiple purification steps to attain required purities, increasing capital and operating expenses. Liquid-liquid chromatography (LLC) has been implemented as an orthogonal purification strategy in the pharmaceutical industry and natural product isolations where it has shown great promise as a scalable and resource-efficient separations technique. As separation challenges emerge in the growing sustainable industrial landscape, LLC has gained attention due to the absence of expensive resins, higher loading capacities compared to other chromatographic techniques, and accessibility to a multitude of unique operation modes. In Multiple Dual Mode (MDM), the liquid nature of the stationary phase is advantageously used to resolve overlapping compounds without changing the identity of the stationary or mobile phases. Understanding and implementing MDM can enhance large scale process model development, however, identifying optimal conditions can be a resource- and time-intensive process. In this presentation, we will discuss the further development of the Cell Utilized Partition (CUP) model for predicting liquid-liquid chromatography processes to accommodate MDM experiments. The CUP model prediction and experimental validation will show the efficient separation of lignin-derived aromatic monomers that are typically difficult to separate using distillation. Finally, a sensitivity analysis using the CUP model will provide efficient process optimization and scale-up strategies.

504c Enhancing Resin Screening with Machine Learning in Protein a Chromatography

Andrea Galeazzi, Steven Sachio, Blaž Likozar, Cleo Kontoravdi, Maria M Papathanasiou

The concept of Quality by Design (QbD) is rapidly reshaping the workflow of how pharmaceutical processes are being developed [1, 2, 3, 4, 5]. The integration of quality restrictions in process design provides a more holistic approach that allows for a significant reduction in research and development time and costs compared to traditional Quality by Testing (QbT) [6, 1]. Quality by Digital Design (QbDD) further improves this by using computer-aided tools for design and optimization [7, 8], leveraging mathematical models for streamlined process development, and faster identification of design spaces. This innovative approach promises significant advancements in sustainable, efficient, and compliant biopharmaceutical manufacturing [9].

A key challenge in QbDD is its reliance on a high-fidelity mechanistic model, i.e. a digital twin [10]. Even though digital twins of the physical assets are of paramount importance, especially from an Industry 4.0 perspective [11], their development can be lengthy and complex. In the case of design variables such as resin type in separations, each alternative would possibly require a tailored digital twin. This can challenge and delay the early stages of process design where multiple variables must be screened and the plethora of candidate designs must be narrowed down. In this context, computer modeling tools can aid and accelerate process development, particularly in cases where the design is multifactorial.

Transferring existing knowledge, particularly in the form of rigorous kinetic modeling, from one system to a different one is a challenging task. Transferring modeling knowledge across systems could result in reducing wet-lab experimentation required and costs, and shaping an evidence-based process development strategy. Nevertheless, digital twins are naturally rather stiff in regards to transferability given that they are customized for a particular unit operation. However, the intrinsic physical knowledge contained in digital twins is invaluable for enabling transfer learning across multiple systems through means of machine learning.

The challenges mentioned above necessitate innovative approaches in process development. In response, this work demonstrates the effectiveness of a data-driven approach, coupled with transfer learning, for design space identification using experimental data. The case study analyzed in this work is a lab-scale Protein A chromatography setup validated on five different resins [12]. In particular, the data generated in the array of experiments has been used for the development of a machine learning algorithm comprising a binary classification neural network to directly estimate the feasible design space. To evaluate the performance of the data-driven model, an accurate digital twin of the unit operation is used and sampled for validation [12].

Transfer learning expands the capabilities of the above-mentioned approach by effectively reducing the need for extensive and costly experimental data while improving predictive accuracy. This is achieved by developing a machine learning model that inherently understands the system's physics through training on diverse rigorous models, capturing their intrinsic characteristics. In practice, the neural network is pre-trained on data generated by digital twins to understand underlying patterns for various resins. The final training is performed solely on experimental data, resulting in an algorithm tailored to the specific experimental campaign.

The presented study highlights the effectiveness of a purely data-driven approach for design space identification and emphasizes the role of transfer learning in improving model performance. Integrating these methods can significantly decrease the need for extensive experimental datasets, optimizing the balance between rigor and computational efficiency. This advancement can be beneficial for biopharmaceutical process development, providing more sustainable, flexible, and cost-effective manufacturing practices.

Acknowledgements
Funding from the UK Engineering & Physical Sciences Research Council (EPSRC) for the i-PREDICT: Integrated adaPtive pRocEss DesIgn and ConTrol (Grant reference: EP/W035006/1) is gratefully acknowledged.

References

[1] A. Simoes, F. Veiga, and C. Vitorino. "Question-Based Review for Pharmaceutical Development: An Enhanced Quality Approach". In: European Journal of Pharmaceutics and Biopharmaceutics 195 (2024).
[2] T.T. Khuat, R. Bassett, E. Otte, A. Grevis-James, and B. Gabrys. "Applications of Machine Learning in Antibody Discovery, Process Development, Manufacturing and Formulation: Current Trends, Challenges, and Opportunities". In: Computers and Chemical Engineering 182 (2024).
[3] R.V. Barenji, R.E. Hariry, D. Demirkol, and T.U. Daim. "Research Landscape Analysis for Quality in Pharma 4.0 Era". In: Technology in Society 76 (2024).0160-791X.
[4] R.S. Nunavath, M.T. Singh, A. Jain, M. Chakma, R. Arivuselvam, and M.S.T.A. Azeeze. "Quality by Design in Pharmaceuticals: A Review of Its Impact on Regulatory Compliance and Product Quality". In: Drug Research 74.1 (2023), pp. 18–23.
[5] Z. Kis, K. Tak, D. Ibrahim, S. Daniel, D. van de Berg, M.M. Papathanasiou, B. Chachuat, C. Kontoravdi, and N. Shah. "Quality by Design and Techno-Economic Modelling of RNA Vaccine Production for Pandemic-Response". In: Computer Aided Chemical Engineering 49 (2022), pp. 2167–2172.50361-4.
[6] G. V. Reklaitis, J. Khinast, and F. Muzzio. "Pharmaceutical Engineering Science—New Approaches to Pharmaceutical Development and Manufacturing". In: Chemical Engineering Science. Pharmaceutical Engineering Science, A Key for Tomorrow's Drugs 65.21 (Nov. 2010), pp. iv–vii.
[7] H. Narayanan, F. Dingfelder, I. Condado Morales, B. Patel, K.E. Heding, J.R. Bjelke, T. Egebjerg, A. Buttè, M. Sokolov, N. Lorenzen, and P. Arosio. "Design of Biopharmaceutical Formulations Accelerated by Machine Learning". In: Molecular Pharmaceutics 18.10 (2021), pp. 3843–3853.1c00469.
[8] Steven Sachio, Cleo Kontoravdi, and Maria M. Papathanasiou. "A Model-Based Approach towards Accelerated Process Development: A Case Study on Chromatography". In: Chemical Engineering Research and Design 197 (Sept. 2023), pp. 800–820.
[9] H. Narayanan, M.F. Luna, M. von Stosch, M.N. Cruz Bournazou, G. Polotti, M. Morbidelli, A. Buttè, and M. Sokolov. "Bioprocessing in the Digital Age: The Role of Process Models". In: Biotechnology Journal 15.1 (2020).
[10] Chaoying Ding and Marianthi Ierapetritou. "A Novel Framework of Surrogate-Based Feasibility Analysis for Establishing Design Space of Twin-Column Continuous Chromatography". In: International Journal of Pharmaceutics 609 (Nov. 2021), p. 121161.
[11] Tsega Y. Melesse, Valentina Di Pasquale, and Stefano Riemma. "Digital Twin Models in Industrial Operations: A Systematic Literature Review". In: Procedia Manufacturing. International Conference on Industry 4.0 and Smart Manufacturing (ISM 2019) 42 (Jan. 2020), pp. 267–272.084.
[12] Matic Grom, Mirijam Kozorog, Simon Caserman, Andrej Pohar, and Bla ˇz Likozar. "Protein A Affinity Chromatography of Chinese Hamster Ovary (CHO) Cell Culture Broths Containing Biopharmaceutical Monoclonal Antibody (mAb): Experiments and Mechanistic Transport, Binding and Equilibrium Modeling". In: Journal of Chromatography B 1083 (Apr. 2018), pp. 44–56.

504d Development of an Advanced API Manufacturing System

Daniel Gregory, Kaitlin Kay, Justin Turnage, Kimberly Penzer, Frank Gupton, James K. Ferri

Active pharmaceutical ingredients (APIs) are traditionally synthesized *via* batch processing on an industrial scale, enabling the production of +500k metric tons of a given API per year. However, batch processing is a cumbersome, labor-intensive strategy that often involves iterative purification steps, large volumes of waste, and chemical variability between each API batch. Additionally, the cost of manufacturing and necessity of hands-on labor has culminated in complex logistical networks which span across multiple continents, requiring chain-of-custody documentation to ensure API synthesis meets FDA regulations. These inefficiencies have led to supply-chain delays for critical lifesaving drugs, a phenomenon exacerbated by SARS-CoV-2. Recently, a new and emerging advanced manufacturing strategy seeks to employ automated, continuous manufacturing systems for the production of drugs *via* flow chemistry. These Next-Gen systems can be configured for high throughput conversion of raw chemical feedstocks into packaged pharmaceutical drug formulations with on-line spectroscopic validation to ensure product purity.

This talk will highlight VCU's construction of a Next-Gen API synthesis module for the production of albuterol sulfate, a bronchodilator asthma drug found on the ASPR drug shortage list. The project is currently in year two in collaboration with Rutgers University. The continuous end-to-end manufacturing system will consist of Module 1: API synthesis (VCU), and Module 2: API formulation, fill-finish, and product packaging (Rutgers). Key engineering decisions will be highlighted including the incorporation of a tubular flow reactor to enable S_N2 amination, a catalytic packed bed reactor (PBR) for hydrogenation, and a series of purification submodules. Chemometric modeling is enabled by correlating off-line analytical characterization *via* high performance liquid chromatography (HPLC) to on-line nuclear magnetic resonance spectroscopy (^1H NMR) and in-line UV-Vis for purity and concentration evaluation. The final automated system will be monitored using process analytical technologies (PAT) to ensure product quality while facilitating the production of 120-1,200 packaged doses per hour. Ultimately, this Next-Gen strategy will help reduce API manufacturing costs and bolster America's pharmaceutical supply chain resiliency.

504e Machine Learning in Food and Drink Manufacturing

Dimitrios Gerogiorgis

ABSTRACT

Food and beverage industries receive key feedstocks whose composition is subject to geographic and seasonal variability, and rely on factories whose process conditions have limited manipulation margins but must rightfully meet stringent product quality specifications. Unlike chemicals, most of our favourite foods and beverages are highly sensitive and perishable, with relatively small profit margins. Although manufacturing processes (recipes) have been perfected over centuries or even millennia, quantitative understanding is limited. Predictions about the influence of input (feedstock) composition and manufacturing (process) conditions on final food/drink product quality are hazardous, if not impossible, because small changes can result in extreme variations. A slightly warmer fermentation renders beer undrinkable; an imbalance among sugar, lipid (fat) and protein can make chocolate unstable.

Though Food Science cannot provide global composition-structure-quality correlations, Artificial Intelligence/AI can be used to extract valuable process knowledge from factory data. For example, food colloid/pulp/paste thickness (viscosity) is instrumental in computational design and optimisation of key processing units (mixers, extruders): it varies enormously for small composition/temperature changes, but must be accurately estimated.

The case of beer, in particular, has been the focus of several of our papers in the past 8 years (2016-24), offering a sound comparison basis for evaluating model fidelity between published precedents and new PINN approaches. Pursuing PINN modelling can cater to greater complexity, both in terms of plant flowsheet as well as target product structure and chemistry, because ML/PINN tools can efficiently predict complex rheological behaviour (food colloids/pastes), which is itself instrumental in computational design and optimisation of key food processing units (mixers, extruders). Traditional (first-principles) descriptions of these necessitate elaborate, high-fidelity (e.g. CFD) submodels of extreme complexity, with at least two severe drawbacks: (1) cumbersome prerequisite parameter estimation with extreme uncertainty, (2) prohibitively high CPU cost.

The currently mature plantwide modelling technology faces dire challenges for industrial implementation in the Food+Drink Manufacturing sector, a lot more so than in other domains (e.g. fuels/pharmaceuticals), due to the ubiquitous multiphase and multicomponent mixtures. Transport properties (e.g. viscosity) vary over orders of magnitude versus composition and temperature. Such extreme variations perplex design and optimisation of manufacturing processes, by inducing structural and parametric uncertainty in first-principle models. Data-driven modelling, however, requires prohibitively expensive plant-wide instrumentation (the norm in refineries, but not at all in food+drink factories, with profit margins significantly lower).

Artificial Neural Networks/ANN and their representational versatility for process systems studies is known for decades [2]. First-principles knowledge, though (mass-heat-momentum conservation, chemical reactions) is captured via deterministic (ODE/PDE) models, which invariably require laborious parameterisation for each particular process plant.

Physics-Informed Neural Networks/PINN [3-4] combine the best of both worlds: they offer chemistry-compliant NN with proven extrapolation power to revolutionise manufacturing, circumventing parametric estimation uncertainty [5] and enabling efficient process control [6].

Fermentation for ethanol [7] and biopharmaceuticals [8-9] was explored using ML/ANN (not PINN) tools, without embedded first-principles descriptions [3]. Exploring ML/PINN potential for Food+Drink Manufacturing optimisation and control is both possible and timely, hence the topic of this contribution, given tools [6] and datasets from our previous publications available [10-11], with a particular focus on industrial fermentation towards beer production. Correlating feedstock properties with processing conditions and product quality via PINN thus showcases promising technical improvements possible towards process optimisation.

ACKNOWLEDGEMENT

Funding from both the EPSRC (RAPID: ReAl-time Process ModellIng and Diagnostics: Powering Digital Factories, EP/V028618/1) as well as the Royal Society via an International Partnership Grant (2023-25) is gratefully acknowledged.

LITERATURE REFERENCES

[1] Gerogiorgis & Bakalis, Digitalisation of Food+Beverage Manufacturing, *Food & Bioproducts Processing*, **128**: 259-261 (2021).

[2] Lagaris et al., ANNs for solving ODEs and PDEs, *IEEE Transactions on Neural Networks*, **9**(5): 987-1000 (1998).

[3] Karniadakis et al., Physics-informed machine learning, *Nature Reviews Physics*, **3**(6): 422-440 (2021).

[4] Raissi et al., Physics-informed neural networks: A deep learning framework for solving forward+inverse problems involving nonlinear PDEs, *Journal of Computational Physics*, **378**: 686–707 (2019).

[5] Lee et al., Machine learning: Overview of recent progresses and implications for the Process Systems Engineering field, *Computers & Chemical Engineering*, **114:** 111-121 (2018).

[6] Alhajeri, Christofides et al., Physics-informed machine learning modeling for predictive control using noisy data, *Chemical Engineering Research & Design,*
186: 34-49 (2022).

[7] Pereira et al., Hybrid NN modelling and particle swarm optimization for improved ethanol production from cashew apple juice, *Bioprocess & Biosystems Engineering*
44: 329-342 (2021).

[8] Peng et al., The ANN approach based on uniform design to optimize the fed-batch fermentation condition: application to production of iturin-A, *Microbial Cell Factories*, **13**: 54 (2014).

[9] Petsagkourakis et al., Reinforcement learning for batch bioprocess optimization. *Computers & Chemical Engineering*, **133**: 106649 (2020).

[10] Rodman & Gerogiorgis, Multi-objective process optimisation of beer fermentation via dynamic simulation, *Food & Bioproducts Processing*, **100A**: 255-274 (2016).

[11] Rodman & Gerogiorgis, Dynamic optimization of beer fermentation: Sensitivity analysis of attainable performance vs. product flavour constraints, *Computers & Chemical Engineering*, **106**: 582-595 (2017).

504f Beyond First-Principles: Harnessing Time-Series Data for Improved Process Modelling.

Nethraa Kannan, Philip Donnellan

The pharmaceutical industry's path towards Industry 4.0 underscores a pivotal transition to integrating advanced AI-based tools in various research domains, with an emphasis on process control. This paper introduces an exploratory study that contributes towards reducing the reliance on state-space or first-principles knowledge. Such conventional and foundational strategies have their limitations in real-time control implementations owing to their extensive computation needs, or inability to capture non-linearities that are usually observed in process data. With continuous manufacturing now producing abundant, information-dense data from improved sensor techniques, it is vital to obtain insights into any underlying long-term time-dependencies to increase prediction accuracy in a dynamic system.

Proof of concept and model-building was carried out from temperature data obtained using a jacketed static mixer system, which is a critical aspect of the pharmaceutical industry. The experiment introduced an impulse temperature change into one of the inlet streams in the system and consequently monitored the altered trajectory of the jacket and mixer outlet temperatures. The predictive performance of this response trajectory was reviewed using a host of models. The baseline outlet temperature predictions were created using two different methods. The first was a simple first-principles model using a heat exchanger approach for a double pipe system, while the second was a classical statistics based time-series approach such as ARIMA (Autoregressive Integrated Moving Average) and VECM (Vector Error Correction Model). These baseline models were compared against advanced machine learning (ML) and deep learning (DL) techniques, specifically XGBoost and LSTM. A combination of window lengths for historical data and future forecast horizons were analyzed when using AI-ML models to ensure a thorough probing of each model's capabilities and limitations. Additionally, the differences in utilizing raw data and feature-engineered data for the AI-ML models was also examined to answer two important questions. The first, the extent to which feature engineering enhances model accuracy, and secondly, whether the improvement in accuracy, if any, justifies the need for domain-specific expertise in developing engineered features. To achieve this, a mix of physics-informed features and mathematical transformation based features were incorporated into the models and assessed accordingly.

Our initial findings highlight that LSTMs obtained the lowest mean absolute error in the outlet temperature predictions across all test data, outperforming the selected baseline and XGBoost models. On evaluating the predictive performance between the two AI-ML models for a one-step ahead forecast averaged across various history lengths, we found the following. The LSTM predictions for the outlet of the mixer have an improved accuracy of 92.690% over XGBoost. For the jacket outlet, XGBoost only slightly outperformed LSTM with an improvement of 2.425%. However, when longer forecasts such as ten step-ahead forecasts were investigated, it was found that LSTM outperforms XGBoost for both jacket outlet as well as mixer outlet predictions. The improvement in accuracy averaged over the range of history lengths showed that LSTMs improved upon XGBoost's predictions of the jacket outlet and mixer outlet by 25.372% and 81.277% respectively. This observation proves that time-dependent relationships contain valuable information that can provide a more nuanced understanding and control of process dynamics. Moreover, of all the tested engineered features, only one feature contributed towards improving the prediction accuracy. This physics-informed feature was created from the mass flow rate contributions of both inlet streams into the static mixer and improved prediction accuracy by 5%. Although beneficial, the addition of such a feature did not markedly improve on the model built using raw sensor data. Therefore, this supports the argument that incorporating LSTMs into existing process control architecture is advantageous for continuous pharmaceutical manufacturing as it can vastly improve prediction accuracies while also reducing the reliance on process-specific experts.

Building on these insights, ongoing efforts are directed towards applying LSTM models trained on on-line UV-vis spectroscopy data for real-time prediction of reaction kinetics in flow chemistry. Further, our future work will be dedicated towards implementing this model into a Model Predictive Control (MPC) framework for flow control. We hypothesize that such models that incorporate long-term dependencies may result in increased robustness for process control as they have the capability to account for time-dependent changes in the system's environment apart from considering process dynamics. However, this requires further investigation and as a result, is a key research question for our future work.

In conclusion, this foundational study highlights the importance of leveraging expansive time-series based sensor data to achieve improvements in predictive modelling. Its potential integration into existing control architecture thereby ensures a smooth and quick transition from ideation to implementation, making it a promising tool for the pharmaceutical industry.

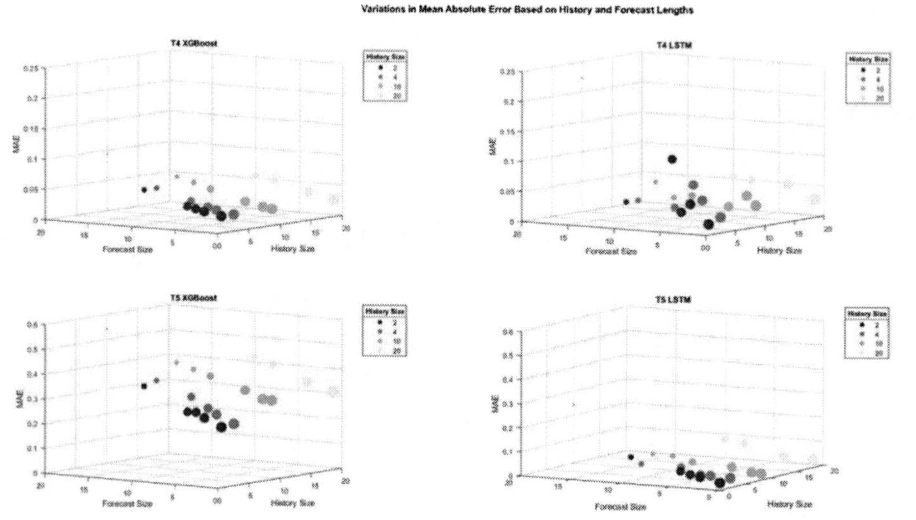

Figure: The MAE values are plotted against the history and forecast size, where T4 is the outlet temperature of the static mixer jacket, and T5 is the outlet temperature of the static mixer. The colours are grouped based on the history size while the size of each marker is based on the forecast length. Shorter forecasts have bigger markers, and longer forecasts have smaller markers. The expected trend is for shorter markers to be higher up on the z-axis which is the case across all four plots, with one outlier for T4 LSTM.

504g Towards Cost-Effective, Sustainable and Flexible Pharmaceutical Supply Chains: Single-Use Vs Stainless Steel Technology

Miriam Sarkis, Alasdair Fyfe, Cleo Kontoravdi, Maria M Papathanasiou

The biopharmaceutical industry is uniquely placed to create value for society, by improving disease outcomes and make ground-breaking therapies available for those in need. As products reach market approval, manufacturers in the sector are pressured to carefully orchestrate and expand resources to meet forecasted demands. In this context, capacity planning activities are typically driven towards manufacturing strategies that decrease therapy costs and maximize product availability. Awareness of environmental sustainability of supply chain operations in life sciences and pharma is becoming increasingly relevant, with recent governmental and industrial initiatives to comply with Net Zero targets by 2030 [1].

Currently, established biopharmaceutical processes may rely on stainless steel equipment which are water and energy intensive [2]. Recent industrial trends for emerging therapeutics and vaccines undergoing scale-up propose instead a shift towards plastics-based equipment to minimize upfront capital investment and help meet global demands faster, which results in a high degree of solid contaminated waste [3]. In this space, computer-aided decision-making can help the benefits and drawbacks of adopting single-use equipment and help compare end-to-end supply chain performance with respect to their cost-effectiveness, environmental impact and scalability.

We present a framework for the quantification and optimization of supply chain sustainability (Figure 1) and illustrate tool capabilities for selected biopharmaceutical products. Firstly, flowsheets are developed to quantify the techno-economic performance of upstream (USP), downstream (DSP) and fill & finish (F&F) process sections of each manufacturing platform. Scale-dependent environmental impacts on midpoint and endpoint indicators are quantified through life cycle assessment (LCA) using ReCiPe 2016. Secondly, a multi-period mixed-integer linear programming (MILP) optimization problem is formulated. The optimization considers a set of candidate network nodes, costs, scale-dependent operational capabilities and demand scenarios and selects candidate supply chain structures, operational plans and suitable process equipment technologies. By fixing network structures and equipment technologies a priori, the optimization can generate a range of candidate solutions which can be compared in terms of (i) supply chain cost, (ii) environmental footprint, (iii) service level and (iv) upfront capital investment.

We consider a range of demand scenarios and biopharmaceutical products to understand product- and process-specific trends. First, the techno-economic analysis and life cycle assessment highlights that, for a given manufacturing scale, the use of single-use technologies in manufacturing may reduce both cost and environmental footprint compared to stainless steel facilities, given the substantial water usage reductions which outweigh the increased plastic waste. Second, the supply chain optimization problem is solved to compare optimal manufacturing configurations with respect to (i) cost, (ii) endpoint indicators vs midpoint indicators, (iii) product service level, through single- and multi-objective optimization methods. Given time-varying demands, the multi-period formulation is solved for fixed manufacturing designs based on stainless steel vs single-use equipment to identify a potential optimal technology switch throughout the scale-up process and the benefit of adopting single-use only in specific process steps.

In this fashion, we demonstrate how the tool can support decision-makers in the sector in developing cost-effective and eco-efficient scale-up strategies that ensure therapy availability from clinical trials to commercialization.

[1] HM Government (2021) Life Sciences Vision

[2] Budzinski K, Constable D, D'Aquila D, Smith P, Madabhushi SR, Whiting A, Costelloe T, Collins M (2022) Streamlined life cycle assessment of single use technologies in biopharmaceutical manufacture. N Biotechnol 68:28–36. https://doi.org/10.1016/j.nbt.2022.01.002

[3] Samaras JJ, Micheletti M, Ding W (2022) Transformation of Biopharmaceutical Manufacturing Through Single-Use Technologies: Current State, Remaining Challenges, and Future Development. https://doi.org/10.1146/annurev-chembioeng

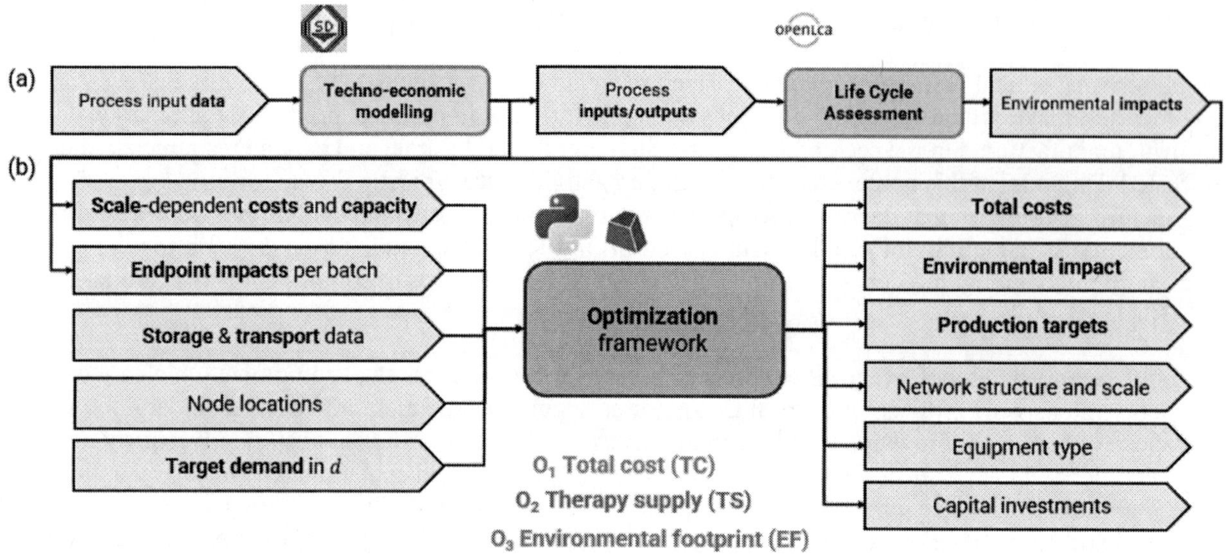

Figure 1: Framework for (a) quantification and (b) optimisation of sustainability of pharmaceutical supply chains

165a The Delicate Interplay between Light, Interfaces and Design:the Complex Dance That Allows 3D Printing to Scale to Manufacturing

Joseph M. DeSimone

The production of polymeric products relies largely on age-old molding techniques. In this talk, I will describe a breakthrough in additive manufacturing—3D printing—referred to as Continuous Liquid Interface Production (CLIP) technology (*Science* **2015**). CLIP, and its recently introduced cousin injection CLIP (iCLIP; *Science Advances* **2022**), embody a convergence of advances in software, hardware, and materials to bring the digital revolution to the design and manufacturing of polymeric products. CLIP uses software-controlled chemistry to produce commercial quality parts rapidly and at scale by capitalizing on the principle of oxygen-inhibited photopolymerization to generate a continual liquid interface of uncured resin between a forming part and a printer's exposure window. Instead of printing layer-by-layer, this allows layerless parts to 'grow' from a pool of resin, formed by light. Compatible with a wide range of polymers, CLIP opens major opportunities for innovative products across diverse industries. Previously unmakeable products are already manufactured at scale with CLIP, including the large-scale production of running shoes by Adidas (Futurecraft 4D); mass-customized football helmets by Riddell; the world's first FDA-approved 3D printed dentures; and numerous parts in automotive, consumer electronics, and medicine. At Stanford, we are pursuing new advances including digital therapeutic devices in pediatric medicine, new multi-materials printing approaches, recyclable materials, and the design of a high-resolution printer to advance technologies in the microelectronics and drug/vaccine delivery areas, including novel microneedle designs as a potent vaccine delivery platform and for the sampling of interstitial fluids for health monitoring and the early detection of disease.

165b Mechanical Properties of 3D Printed Thermoplastics Via Material Extrusion from Surface Topology

Bryan Vogt, Bevan Harbinson

Defects from one layer are known to impact subsequent layers to lead to catastrophic failure of 3D printed parts using filaments. These large scale defects are generally easy to identify, but this raises the question what about minor errors in printed parts. Do minor deviations in the print on one layer lead to a quantifiable change in subsequent layers? Here, we describe how the surface topology of a final part can be used to provide insights into the mechanical properties to describe part-to-part variation even under the same printing conditions. The variance in the ultimate tensile strength (UTS) of the printed parts can be correlated with differences in the surface topology of the printed parts. We demonstrate how use of Fourier transforms provides a simple metric that describes how minor, difficult to visualize changes in the surface topology are related to the UTS. This correlation can be used to identify outlier parts and provide a route for qualification of additively manufactured parts without necessitating overdesign to address variance in the printing process. This post-print characterization does not require significant investment for each tool, nor their modification, but provides an opportunity to provide value through improving reliability. If time permits, extension of these concepts to other mechanical properties using vision transformer (ViT) for enhanced image analysis will be discussed.

165c 3D Printing Biohybrid Polymer Composites

Samson Adelani, Olivia Pear, R. Konane Bay

The majority of 3D-printed biohybrid polymeric materials are printed on flat surfaces limiting the complexity and size of the materials. However, since these inks of materials are typically extremely soft, they require crosslinking as they are 3D printed. Therefore, fixing the final mechanical properties of these materials during 3D printing. We overcome these challenges, by utilizing a 3D support media that fixes the ink into place allowing for the cells to grow prior to post-processing (biomineralization or crosslinking). Here, we 3D print the bacteria, *Escherichia coli* or mycelium, *Aspergillus niger* with crosslinkable polymer into a granular hydrogel support bath. We demonstrate how crosslinking and biomineralization post-incubation of the 3D printed materials impact the mechanics of the biohybrid polymer composites. Our results thus provide fundamental insights into how the growth of cells in confined environments can affect the mechanics of 3D-printed microbial polymer composites.

165d *Directed Self-Assembly of Nanostructured Polymers Via 3D Printing: achieving Mechanically Tailored Soft Architectures*

Emily Davidson

Nanostructured polymers, such as liquid crystalline polymers and block copolymers with intrinsically anisotropic self-assembled nanostructures, can be macroscopically oriented to exhibit multiple orders of magnitude in anisotropy in their material properties and stimuli response. Here, we leverage material extrusion additive manufacturing to crease soft hierarchically ordered structures, bridging the interface between top-down manufacturing and bottom-up self-assembly. We demonstrate that in both classes of materials we can control the extent of nanostructure alignment and functional anisotropy via the flow the material undergoes during 3DP; for thermoplastic elastomer block copolymers, post-printing thermal annealing additionally plays a critical role in driving long-range alignment. To demonstrate the versatility of this approach, we tailor print paths to fabricate soft architectures with designer macroscopic behavior such as controlled localization of tensile and flexural strain upon deformation.

165e Design and Fabrication of Conductive 3D Printed Carbon Lattices Via Gel Infusion and Polymerization of Acrylonitrile and Subsequent Pyrolysis

Philip Onffroy, Max A. Saccone, Dennis Schroeder, Jacob Dobson, Montana Lagat, Jason M. Kronenfeld, Samuel Chiovoloni, Jennifer Q. Lu, Joseph M. DeSimone

High-temperature pyrolysis of 3D printed polymers to produce conductive pyrolytic carbon is gaining prominence in sustainable energy storage, carbon capture, and electrification applications. Pyrolysis necessitates structures with high surface-to-volume ratios to maintain mechanical integrity, as thinner features allow for the efficient escape of volatile components during heating, thereby preventing deformation. However, traditional molding methods are not able to fabricate large objects with microscale features, such as beam-based lattice structures. High resolution vat photopolymerization (VP) 3D-printing of non-moldable structures with high surface area to volume ratios, such as via the state-of-the-art continuous liquid interface production (CLIP) method, therefore adds significant value to pyrolysis. One of the most widely used materials in pyrolysis is polyacrylonitrile (PAN), which is a precursor for generating conductive high-strength carbon fibers that are commonly used in lightweight structural applications and have been incorporated into battery systems for sustainable energy storage. However, limited literature exists that demonstrates generation of high-resolution structures (< 100 μm features) composed of PAN by a scalable VP 3D-printing method, such CLIP. During VP 3D-printing, the formation of a crosslinked network, which is crucial for preserving high-resolution architectures during photo-initiated polymerization, is hindered by the linear polymer nature of PAN which leads to precipitation reactions and subsequent powder formation.

To overcome the challenges faced by VP 3D-printing high-resolution PAN structures for pyrolysis, we have developed a novel method for producing 3D carbon structures from PAN. This process involves: (1) 3D printing a microscale crosslinked acrylate gel lattice sacrificial scaffold via CLIP, (2) infusing the gel scaffold structure with acrylonitrile monomer and a thermal initiator, (3) polymerizing the AN monomer to PAN linear chains inside the gel scaffold structure forming a semi-interpenetrating polymer network, and then (4) subjecting the PAN-infused gel lattice structure to thermal treatment processes (e.g. an oxidative pre-treatment, isothermal holds, and pyrolysis) that converts the PAN to 3D pyrolytic carbon (Figure 1). This method is inspired by previous work by Saccone, M. and Greer, J. *et al.* investigating metal hydrogel infusion.

We have investigated the impact of pyrolysis conditions on the char yield of these PAN-swelled gel scaffolds over a wide range of temperatures and excipient gas (air, nitrogen) flow rates. With optimized conditions, we demonstrate > 40% char yield by mass, which is comparable to industry values of char yield for conventionally electrospun PAN fibers. Characterization of the resulting pyrolytic carbon structures has demonstrated that they exhibit evidence of graphitization, they are electrically conductive, and they exhibit promising voltage potential ranges for energy storage. This method additionally shows promise for incorporation of other high char yield linear polymers beyond PAN into high-resolution lattice structures to tailor pyrolytic carbon properties. PAN-based pyrolytic carbon lattice structures developed via our novel gel infusion method hold significant promise for use in commercial applications including electrodes for battery energy storage, susceptors for electrifying thermochemical reactors, and substrates for carbon capture electrocatalysis.

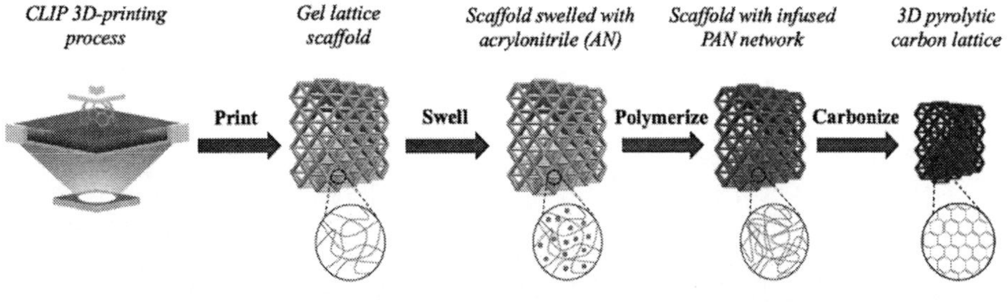

Figure 1. Polyacrylonitrile (PAN) gel infusion in a CLIP 3D printed sacrificial lattice scaffold followed by pyrolysis

165f Molecular Design of Polymeric Powder Bed Fusion Feedstocks to Tailor Properties of Printed Parts By Enhanced Coalescence

Akan George, Jackson Bryant, Michael Bortner, Christopher B. Williams, Mark Dadmun

Coalescence is a fundamental aspect of 3D printing that directly impacts the strength, quality, and properties of printed parts. In this presentation, we will discuss our recent work that utilizes a processing protocol based on liquid-liquid phase separation to controllably create polypropylene (PP) powders from 12k, 250k, and 340k PP molecular weight (Mw) along with their blends and tune their coalescence dynamics tailored for powder bed fusion. Examination of the particle coalescence of these powders shows that the addition of 12k significantly increases the coalescence rate of high Mw PP, consequently reducing the void space and increasing the mechanical strength of the resultant printed parts. The coalescence of these multi-components powders follows the Hopper model, offering insight into the underlying mechanisms governing particle consolidation. A holistic interpretation of the data shows that the 12k PP does not surface segregate in the powder but is homogeneously distributed, improving the coalescence process by lowering the extensional viscosity of the polymer melt. These findings provide a straightforward mechanism to molecularly design powder coalescence properties and offer avenues to broaden the feedstocks available for powder bed fusion.

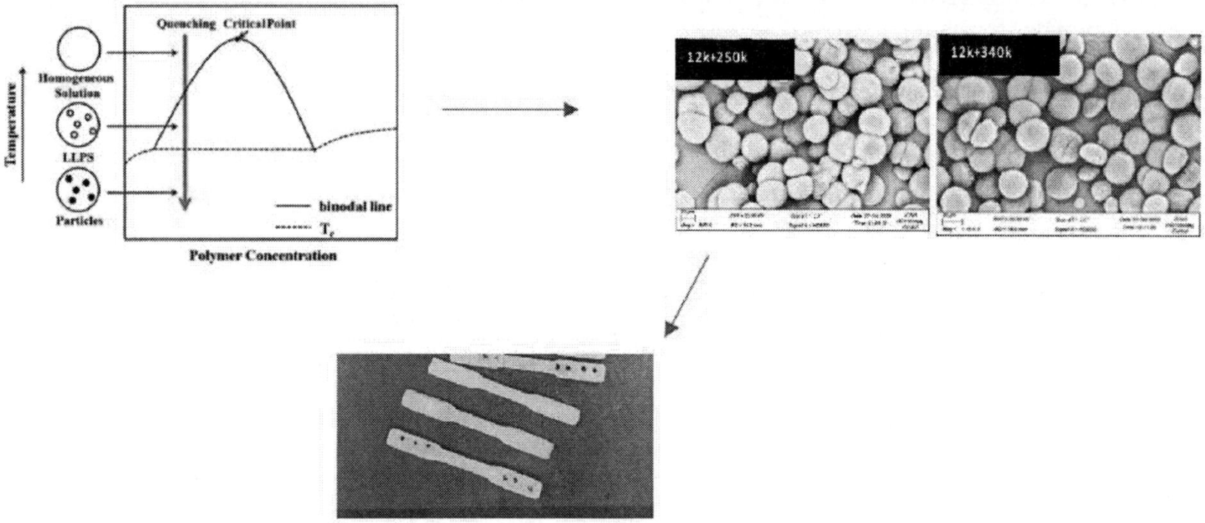

235a Ultra-Tough Ionogels for 3D Printing

Michael Dickey

Ionogels – polymer networks containing ionic liquids—are compelling materials for energy storage devices, ionotronics, and actuators due to their excellent ionic conductivity, thermal and electrochemical stability and nonvolatility. However, most existing ionogels suffer from low strength and toughness. Here, we report a simple one-step method to achieve ultra-tough and stretchable ionogels that is compatible with 3D printing. Randomly copolymerizing two common monomers in ionic liquid leads to a single covalent network with controllable polymer- and solvent-rich phases that form in situ due to the phase behavior of the polymer in ionic liquid. The polymer-rich phase forms hydrogen bonds that dissipate energy and thereby toughen the ionogel during extension, while the solvent-rich phase remains elastic to enable large strain. The copolymer ionogels composed of acrylamide and acrylic acid exhibit extraordinary mechanical properties, including fracture strength (12.6 MPa), fracture energy (~24 kJ m-2), and Young's modulus (46.5 MPa), setting new records among reported ionogels. The tough ionogels are highly stretchable (~600% strain) and possess good self-recovery, as well as excellent self-healing and shape-memory properties. This concept extends to other monomers and ionic liquids, which offers a promising and general way to tune microstructure in situ during one-step polymerization that solves the longstanding mechanical challenges in ionogels.

235b Effects of ABS Variants with Distinct Microstructures on Interlayer Bond Formation in Matex AM Based on Their Morphological, and Rheological Aspects.

Juhyeong Lee, Nikhil Patil, Jay Park

Acrylonitrile-butadiene-styrene (ABS) is an amorphous polymer, generally consisting of a continuous Styrene-Acrylonitrile (SAN) copolymer phase in which rubber spherical formed particles (mainly polybutadiene-PB) grafted with SAN are dispersed. This microstructure is mainly influenced by various factors such as PB rubber particle size and distribution, and others. The resulting microstructure allows ABS to lead to a complex failure mechanism involving cavitation, crazing, and localized shear yielding, resulting in a combination of stiffness and toughness.

Due to these favorable mechanical properties and its rheological behavior suitable for printing, ABS is one of the most common materials used for material extrusion additive manufacturing (MatEx AM). Despite the availability of a wide range of ABS grades on the market and numerous studies utilizing them, there has been limited research into understanding the relationships between ABS microstructure and interlayer bond formation.

In this study, we evaluate these relationships by comparing the as-printed and annealed impact strengths of various ABS grades with distinct morphological and rheological characteristics. Notably, ABS grade with larger rubber particle size exhibits a remarkable annealing effect. Despite initially having the lowest as-printed impact strength (1,396 J/m2), this grade achieves the highest impact strength (17,479 J/m2) after fixtured annealing, closely approaching the performance of their bulk injection-molded counterparts.

We hypothesize that the unique microstructure of ABS inhibits interlayer bond formation during printing due to its unfavorable rheological properties. However, our results suggest that after annealing, these initially unfavorable rheological properties induce increased interlayer contact pressure and potentially promote a more even distribution of rubber particles within the interlayer. Our hypothesis is supported by morphological analysis using AFM modulus mapping (e.g., rubber size and distribution on weld line) and rheological analysis using parallel plate rheometer (e.g., reptation time or Weissenberg number).

235c Formulating Biomolecular Gels for 3D Bioprinting Using High-Throughput Autonomous Microrheology

Rhett L. Martineau, Alexandra V. Bayles, Maneesh K. Gupta, Matthew Helgeson

Formulating inks and resins for 3D bioprinting of cell-laden polymer gels presents a complex rheological optimization. If gelation kinetics are too slow, printability and 3D cell encapsulation can be compromised due to long solidification times; if gelation is too fast, cell viability is lost due to incompatibly high crosslinking densities. Finding an optimal "Goldilocks zone" can therefore require rheological screening of large formulation spaces of gelation kinetics, with material requirements that are prohibitive for precious materials including novel polymer chemistries and biomolecules. To address this challenge, we have developed a high-throughput, low-volume, fully autonomous rheology platform for polymer formulation discovery and design. Based on passive probe microrheology, the workflow combines robotic sample preparation, automated microscopy, custom-built high-throughput data analysis and ML-guided active learning to achieve a fully autonomous workflow that can be used to screen large formulation spaces with complex objective functions and minimal time and material requirements. As a demonstration, we apply the workflow to bioinks comprised of crosslinkable protein and peptide-synthetic hybrid materials, in which the optimization of gelation kinetics involves high-dimensional compositional search spaces. We show that achieving optimal gel times from this workflow can be achieved autonomously in less than 24 hours and with less than one milligram of polymer material. The resulting ink formulations exhibit acceptable cell encapsulation and viability without any additional optimization. These results therefore show considerable promise for using high-throughput autonomous microrheology to optimize formulations for bioprinting, and polymer additive manufacturing more broadly.

235d High-Resolution Stereolithography: Negative Spaces Enabled By Control of Fluid Mechanics

Ian Coates, William Pan, Max Saccone, Gabriel Lipkowitz, Eric Shaqfeh, Joseph M. DeSimone

Stereolithography enables the fabrication of three-dimensional (3D) freeform structures via light-induced polymerization. However, the accumulation of UV dose within resin trapped in negative spaces, such as microfluidic channels or voids, can result in the unintended closing, referred to as overcuring, of these negative spaces. We report the use of injection continuous liquid interface production (iCLIP) to continuously displace resin at risk of overcuring in negative spaces created in previous layers with fresh resin to mitigate the loss of Z axis resolution. We demonstrate the ability to resolve 50 mm microchannels, breaking the historical relationship between resin properties and negative space resolution. Using the iCLIP approach, we fabricated proof-of-concept 3D free-form microfluidic devices with improved design freedom over device material selection and resulting properties for applications ranging from microneedle drug delivery to advances separations.

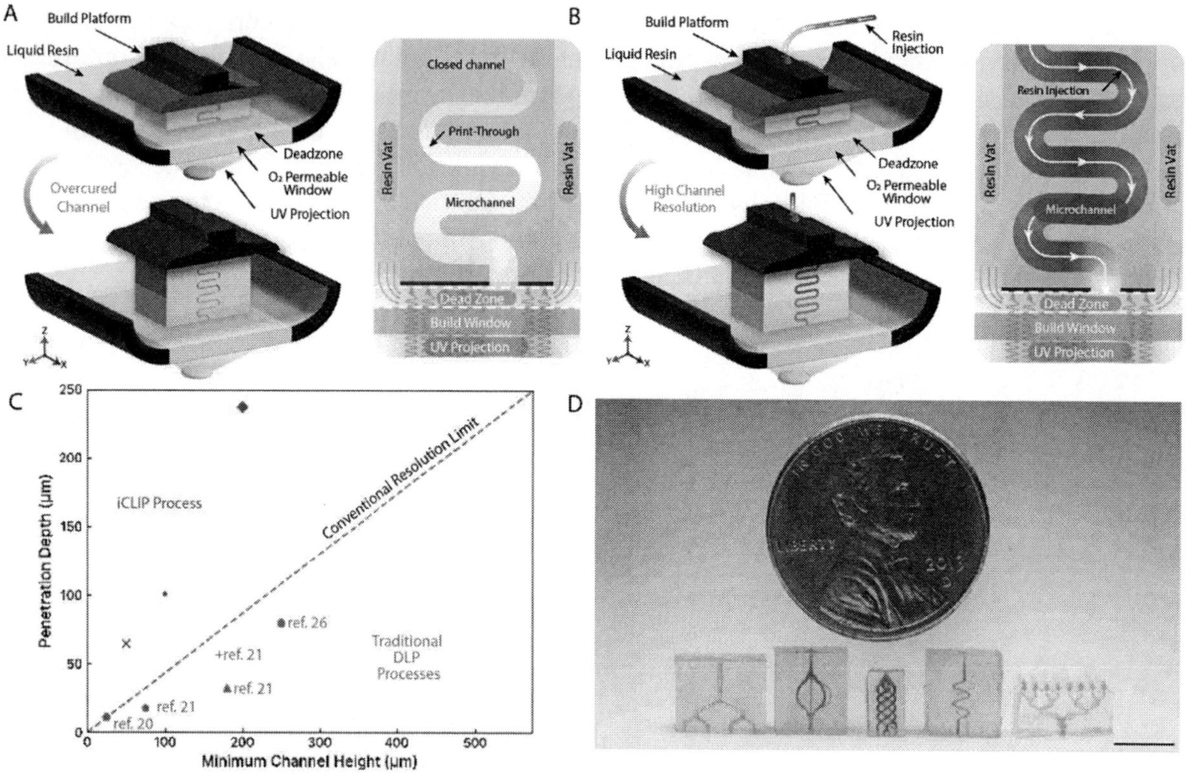

235e Selective Laser Sintering of Polymer Nanocomposite Powders for Precise Fabrication of 3D Hierarchical Composites

Weinan Xu

Selective laser sintering (SLS) (also named powder bed fusion) has been regarded as the most promising polymer 3D printing technology for many industrial applications. Its unique advantages compared with extrusion- or photopolymerization-based technologies include the ability to fabricate complex geometries without support structures, highly isotropic properties of the printed parts, and batch production of multiple parts in one printing. However, the major limitation of SLS is the limited polymer powder selection (95% is based on polyamide) and difficulty in multi-material printing. Here, I will present our recent progress in the development of new types of polymer nanocomposite powders for SLS and the fabrication of hierarchical composite 3D structures. For instance, we successfully developed a strategy for SLS of polyethylene and its composites with excellent printability and highly tunable internal structures. We also demonstrated that the SLS printing and simultaneous laser-induced graphitization of polyimide microparticles into 3D hierarchical graphene structures with precisely controlled geometry and internal structure. 3D graphene-metal oxide structures can also be fabricated by this strategy, which have superior electrochemical properties for energy storage and catalysis applications.

235f Predicting the 3D Printability of Polymer-Based Ink Using Machine Learning.

Md Shakibur Rahaman Chowdhury, Brandon Bell, Trevor Melton, Amitabha Dey, Asmaa Alawbali, Mohammad Azad

3D printing enhances the utilization of materials more efficiently than conventional manufacturing methods and facilitates the creation of intricate designs customized to meet specific requirements. Additionally, it enables the quick production of prototypes, expediting the development of new products. Nevertheless, the creation of 3D-printed items necessitates much experimentation and refinement. The lack of knowledge regarding the properties of the polymer material used in ink production and its subsequent influence on the result of 3D printing is the cause of this trial-and-error effort. As a result, we need to change the process parameters for successful printing if we change the material in the ink preparation. This work aims to establish a machine learning model framework capable of accurately predicting the outcome of extrusion-based 3D printing using polymer-based ink for high-value product (i.e., pharmaceuticals) applications. The goal is to reduce the need for trial and error, eliminate material waste, and ultimately lower costs. As a demonstration of feasibility, we formulated 24 distinct ink samples using a range of material compositions. We then conducted 220 observations to analyze the effects of changes in the process parameters. We conducted three fundamental rheology tests on each ink and analyzed their properties in terms of pre-printing stage, in-process condition, and post-printing stage. The rheology test findings are converted into data and used as input for the model. Next, we input the data into machine learning (ML) models (Random Forest and Artificial Neural Network) incorporating process variables and assess the performance of each ML model in terms of predictability. The results of our study demonstrate significant progress in 3D printing prediction, with accuracy rates of 71.4% and 63.3%, respectively. These findings indicate a high level of accuracy and successful prediction. The F1 score for Random Forest and Artificial Neural Network is 79% and 76%, respectively, validating our proof of concept using limited experimental data.

AUTHOR INDEX